Companion Animal Anatomy
A Photo Atlas

Heather W. Dunn

Department of Animal & Veterinary Science
Clemson University

PREFACE/ ACKNOWLEDGEMENTS

Companion Animal Anatomy: A Photo Atlas is designed for the student seeking a career as a veterinary technician or in the field of veterinary science. The aim of this photographic collection is to illustrate companion animals as they would be presented to the health care provider.

Anatomical figures of canine and feline species are presented in this full color collection of illustrations and photographs. Most chapters begin with images of histological tissue represented by the corresponding organ system. A glossary is included as supplementary information, and the atlas concludes with perforated review pages from each chapter for reinforcement of material.

During all phases of photography, animals were never sacrificed for the sole purpose of collecting images. Fresh tissue dissections were performed on animals that died of natural causes or they were euthanized for failing health. Images of live animals were taken during routine handling procedures at Tiger Town Animal Hospital in Clemson SC. Euthanized dogs and cats were provided by Oconee County Humane Society in Seneca, SC.

This photo atlas is the result of a joint project between the Department of Animal & Veterinary Sciences (AVS) and Creative Inquiry at Clemson University in Clemson, SC. Ms. Kaitlin Iulo provided illustrations. Students Ms. Megan Kelley and Ms. Tina Rowland were responsible for compiling hundreds of images, organization, layout, design, and revisions. Megan and Tina devoted hours of their time including evenings and weekends to complete this work. Thank you ladies for providing your talent, knowledge, energy and humor. I truly could not have completed this project without your assistance.

Dr. Jim Strickland, Department Chair of AVS has promoted undergraduate involvement with this project and provided support to continue developing new Creative Inquiry projects. The staff at Godley Snell Research Center (GSRC), under the guidance of Dr. John Parrish DVM has provided facilities for all necropsy. Thank you GSRC for the use of your beautiful facility.

And finally, I would like to thank Dr. Nathan J. Craddock DVM and his staff at Tiger Town Animal Hospital for allowing us to photograph companion animals during surgeries and routine care. Dr. Craddock kindly provided radiograph images compiled in this photo atlas. We are grateful for your support of the Clemson University AVS program and for being such a caring, compassionate veterinarian.

The author welcomes feedback, suggestions and comments regarding the material covered in this photo atlas. Please send comments to the following address:

Dr. Heather W. Dunn
Department of Animal & Veterinary Sciences
138 Poole Agriculture Building
Clemson University
Clemson, SC 29634
walkerd@clemson.edu
864-656-2516

Table of Contents

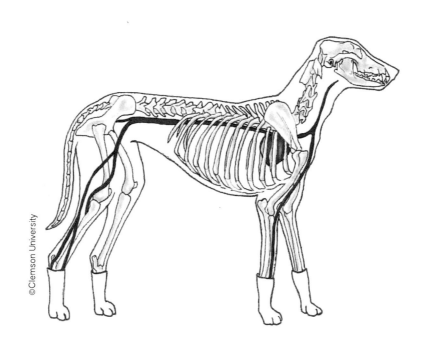

©Clemson University

Chapter 1

Introduction to Anatomy

©Clemson University

Limited examples of each canine group.

Herding Group

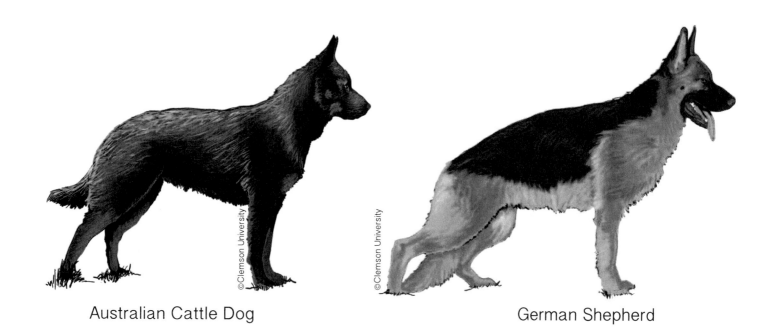

Australian Cattle Dog

German Shepherd

Hound Group

Dachshund

Greyhound

Canine - GROUPS & BREEDS

Limited examples of each canine group.

Sporting Group

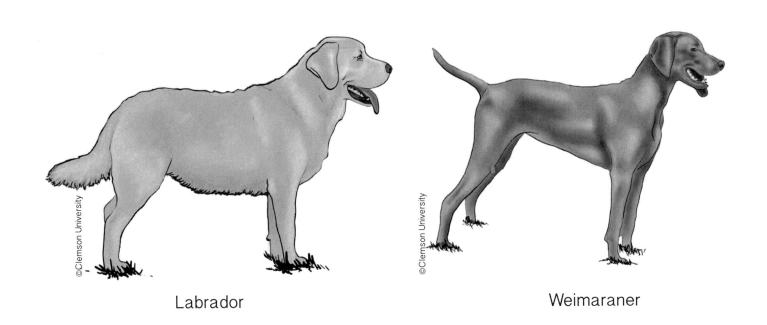

Labrador

Weimaraner

Terrier Group

Staffordshire Bull Terrier

Scottish Terrier

Limited examples of each canine group.

Toy Group

Chihuahua

Shih Tzu

Working Group

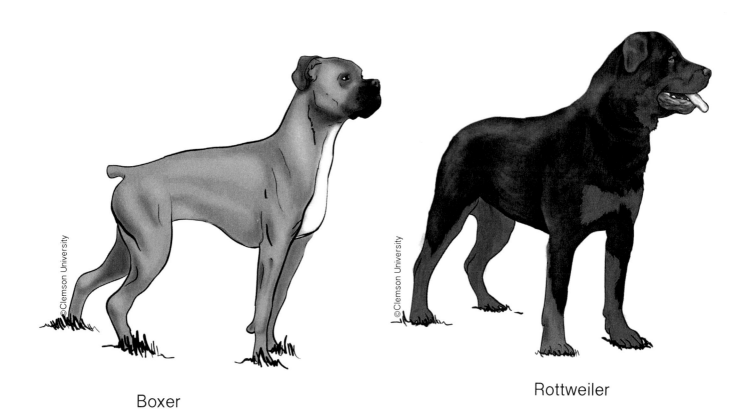

Boxer

Rottweiler

Head Morphologies

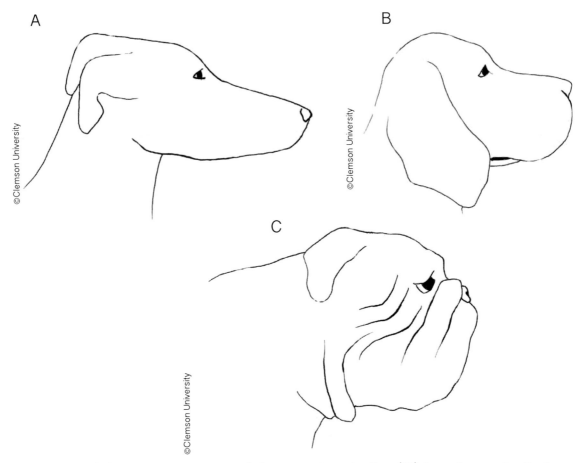

Representatives of (A) dolichocephalic, (B) mesaticephalic, (C) brachycephalic breeds

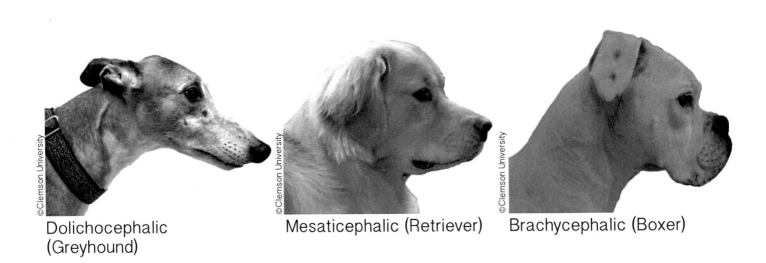

Dolichocephalic
(Greyhound)

Mesaticephalic (Retriever)

Brachycephalic (Boxer)

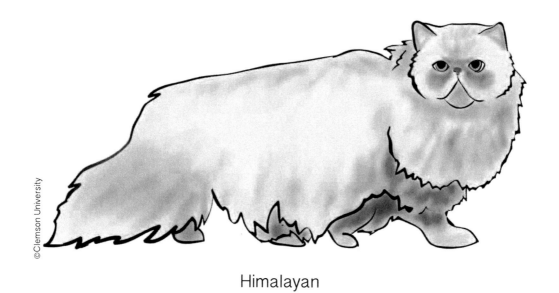

Himalayan

Siamese

Bengal

CANINE - Directional Planes

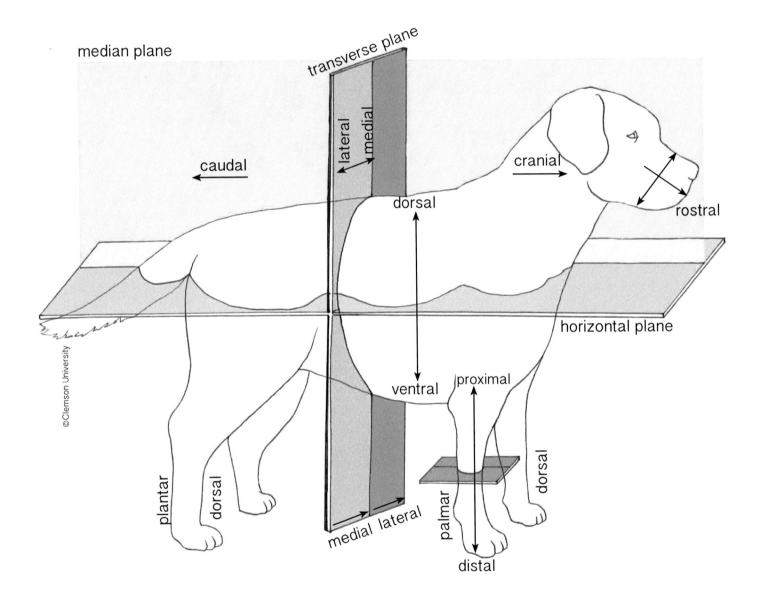

median plane

transverse plane

lateral

medial

caudal

cranial

dorsal

rostral

horizontal plane

©Clemson University

ventral

proximal

plantar

dorsal

medial   lateral

palmar

dorsal

distal

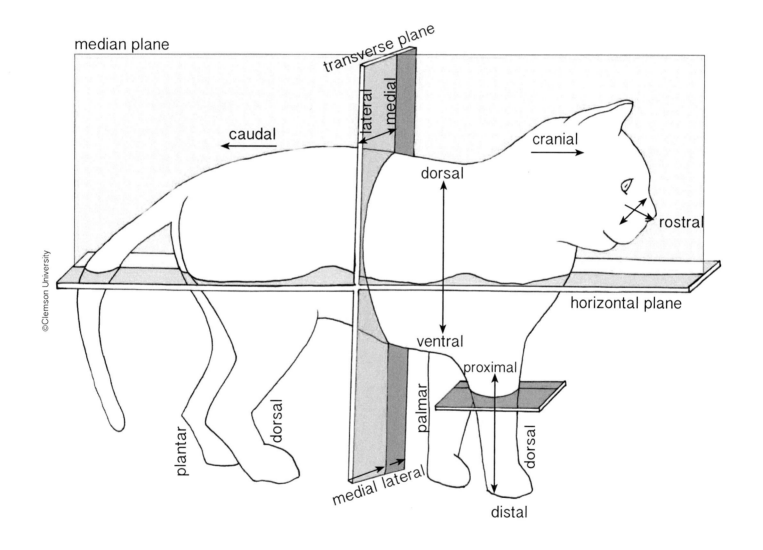

CANINE - General Anatomy

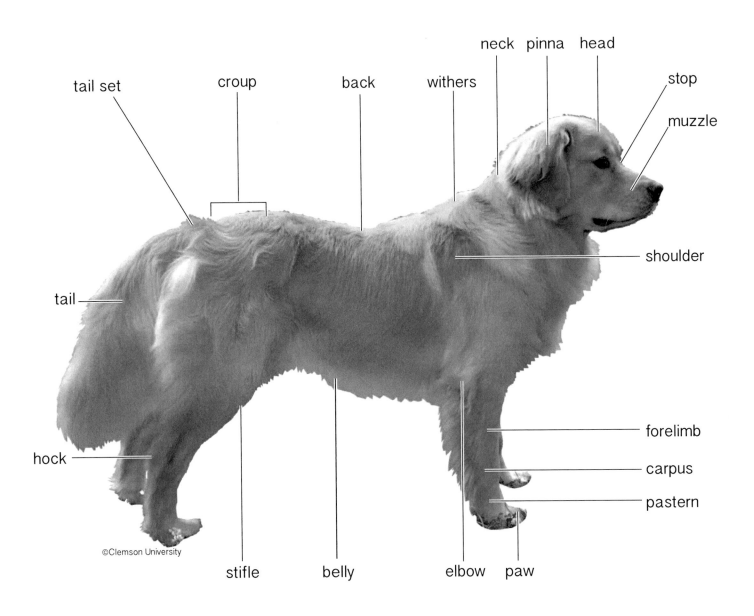

neck pinna head

tail set   croup   back   withers   stop

muzzle

shoulder

tail

forelimb

carpus

pastern

hock

©Clemson University

stifle   belly   elbow paw

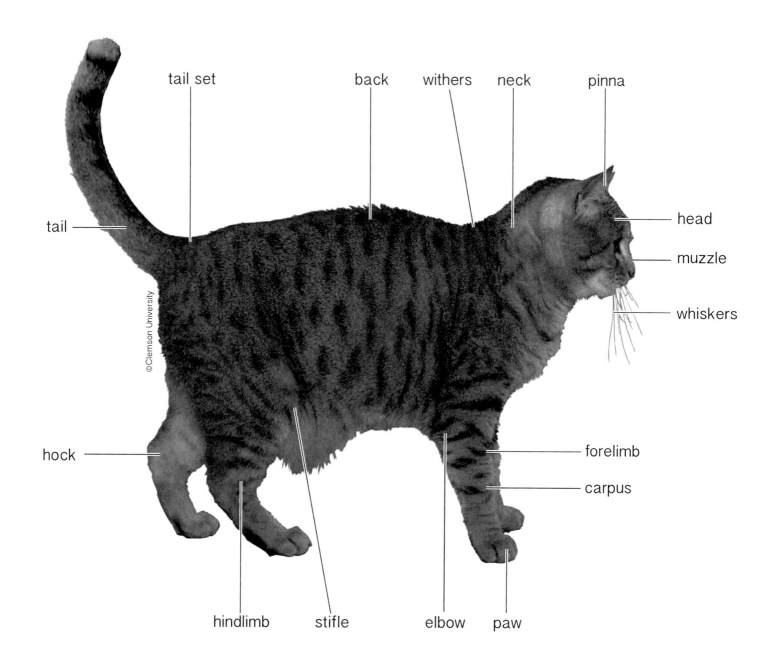

tail set   back   withers   neck   pinna

tail

head

muzzle

whiskers

©Clemson University

hock

forelimb

carpus

hindlimb   stifle   elbow   paw

DORSAL VERTEBRAL REGIONS

Canine

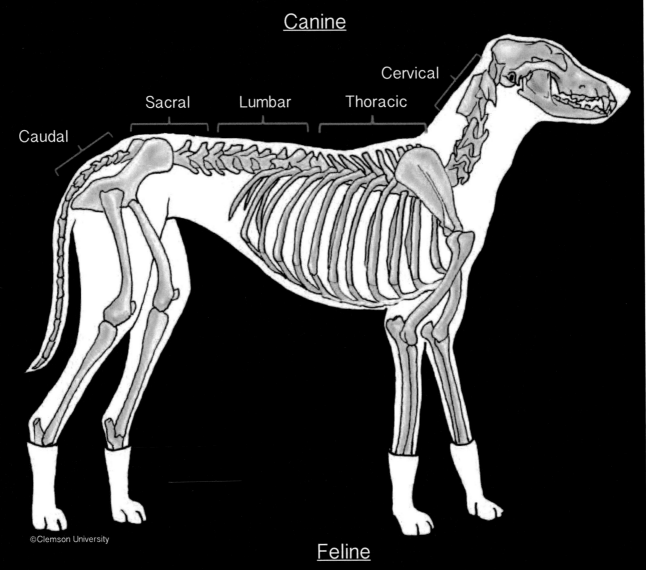

Cervical

Sacral   Lumbar   Thoracic

Caudal

©Clemson University

Feline

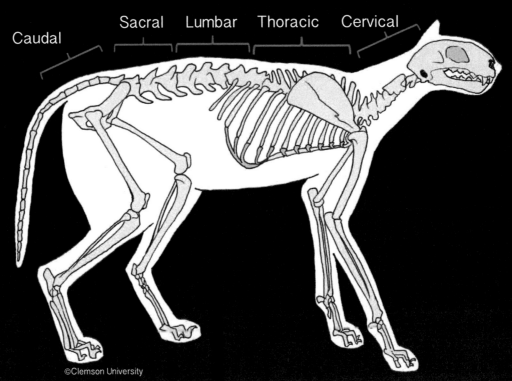

Sacral   Lumbar   Thoracic   Cervical

Caudal

©Clemson University

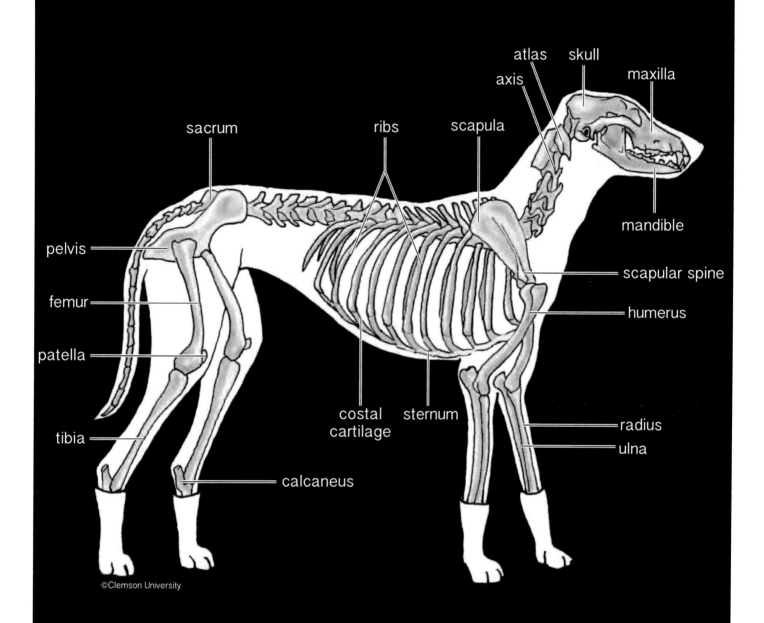

atlas skull

axis maxilla

sacrum ribs scapula

mandible

pelvis

scapular spine

femur

humerus

patella

tibia

costal cartilage sternum

radius

ulna

calcaneus

©Clemson University

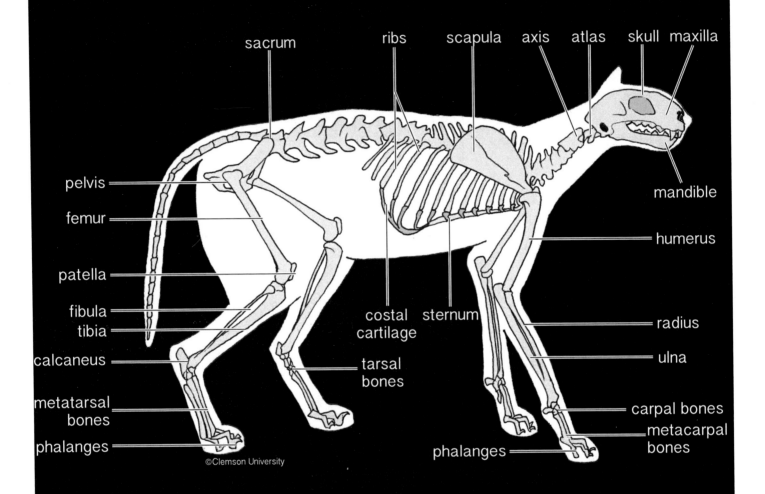

sacrum · ribs · scapula · axis · atlas · skull · maxilla · mandible · humerus · radius · ulna · carpal bones · metacarpal bones · phalanges · pelvis · femur · patella · fibula · tibia · calcaneus · metatarsal bones · phalanges · costal cartilage · sternum · tarsal bones

©Clemson University

Integumentary System

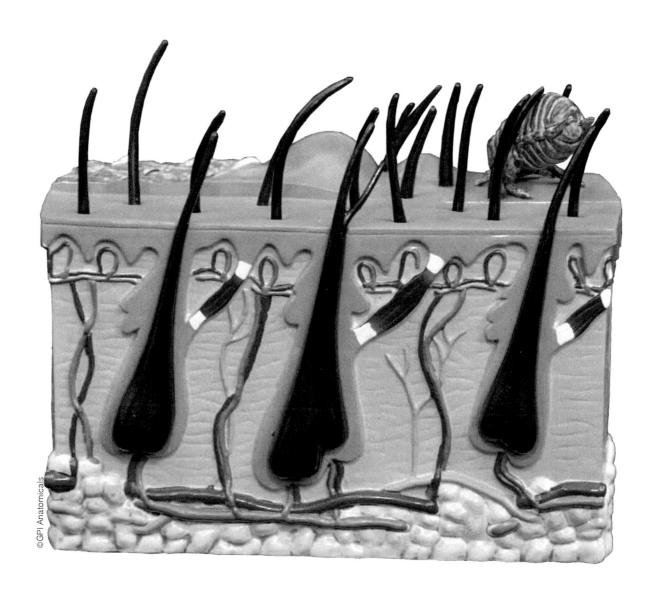

Histology - SKIN

Glabrous Skin (Nose)

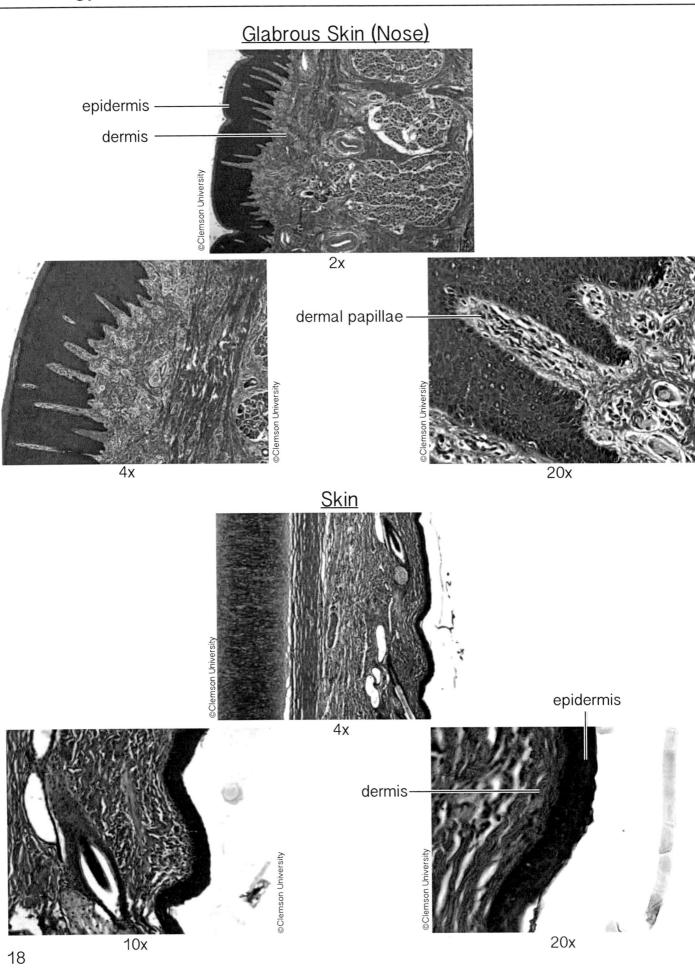

epidermis —

dermis —

©Clemson University

2x

4x

©Clemson University

dermal papillae —

©Clemson University

20x

Skin

©Clemson University

4x

©Clemson University

10x

©Clemson University

epidermis

dermis —

©Clemson University

20x

Canine Normal Skin

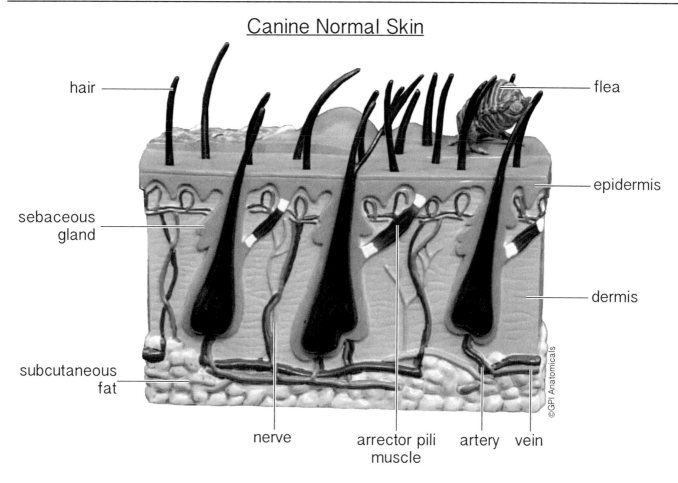

hair

flea

epidermis

sebaceous gland

dermis

subcutaneous fat

©GPI Anatomicals

nerve

arrector pili muscle

artery

vein

Canine Inflammatory Response to Flea Bite

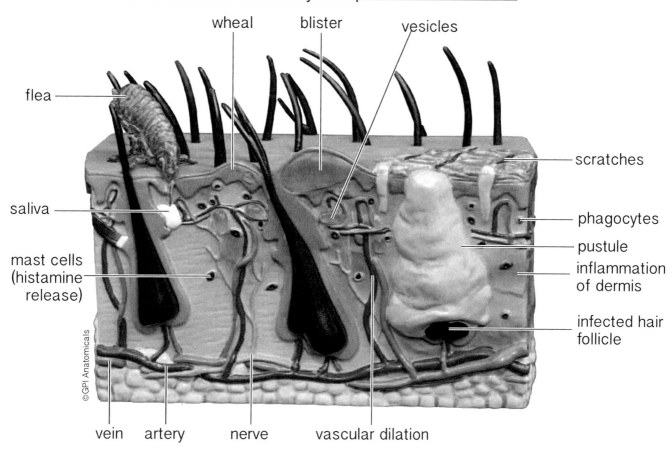

wheal

blister

vesicles

flea

scratches

saliva

phagocytes

mast cells (histamine release)

pustule

inflammation of dermis

infected hair follicle

©GPI Anatomicals

vein

artery

nerve

vascular dilation

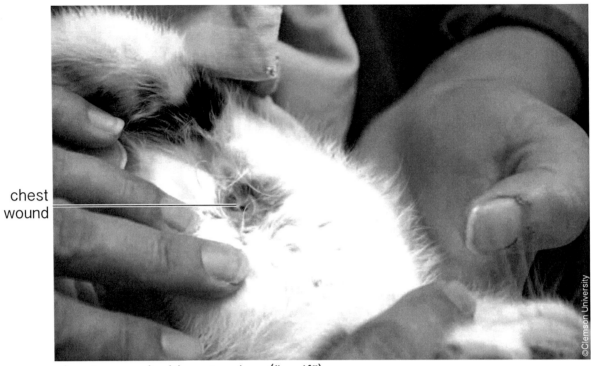

chest
wound

chest wound with cuterebra ("wolf") worm

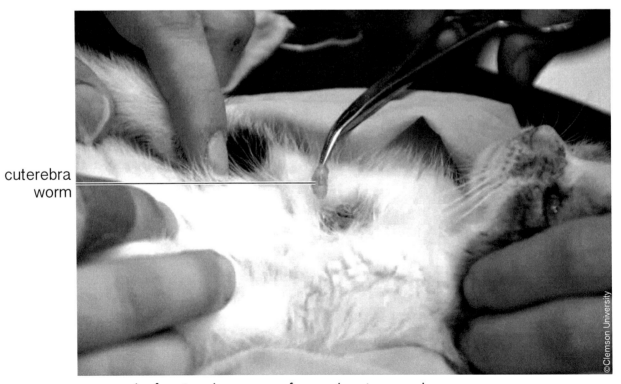

cuterebra
worm

removal of cuterebra worm from chest wound

Canine

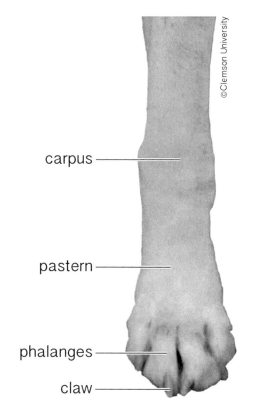

carpus

pastern

phalanges

claw

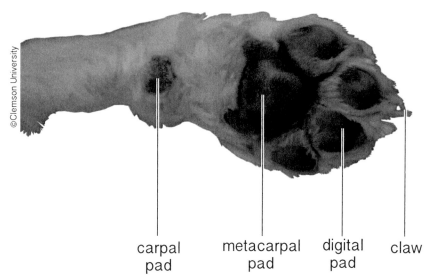

carpal pad

metacarpal pad

digital pad

claw

Feline

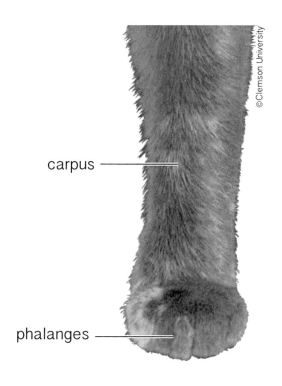

carpus

phalanges

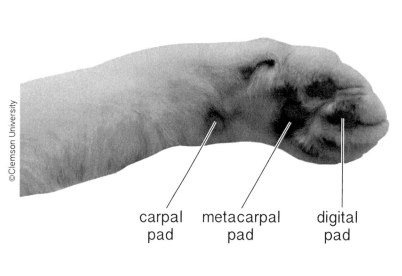

carpal pad

metacarpal pad

digital pad

CLAWS

Canine

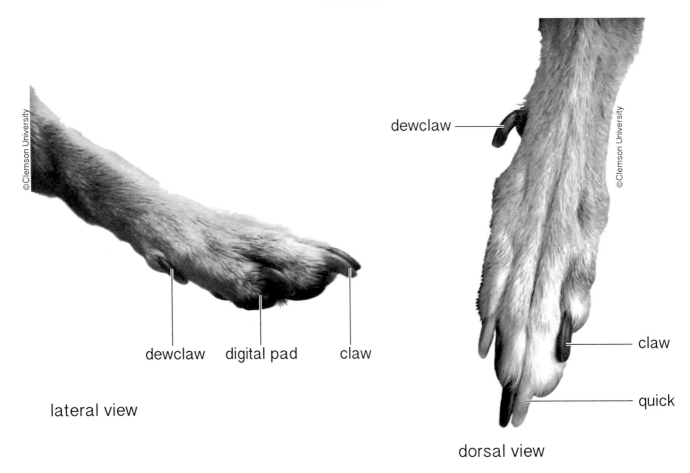

dewclaw — digital pad — claw

lateral view

dewclaw —

claw —

quick —

dorsal view

Feline

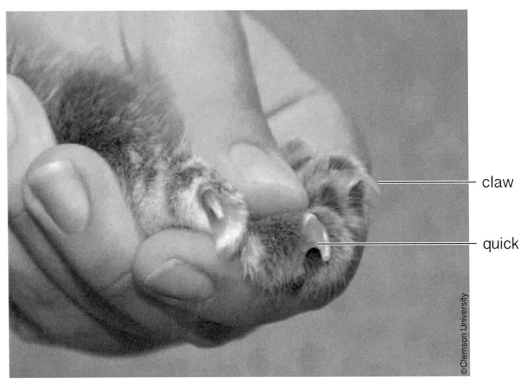

claw —

quick —

extended claws

Chapter 3

Skeletal System

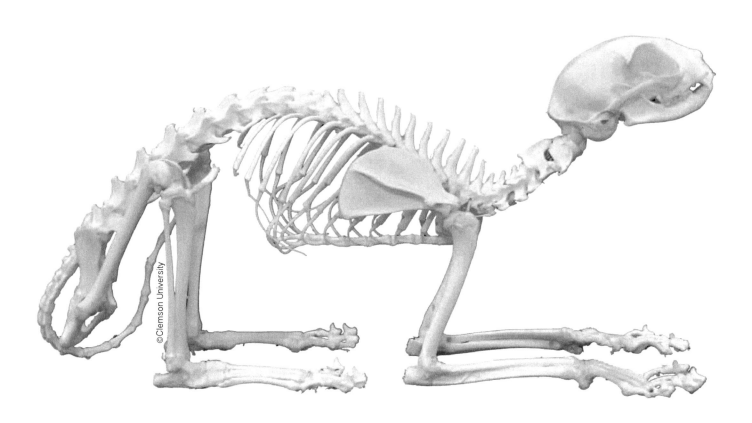

©Clemson University

Histology - SKELETAL SYSTEM

Bone

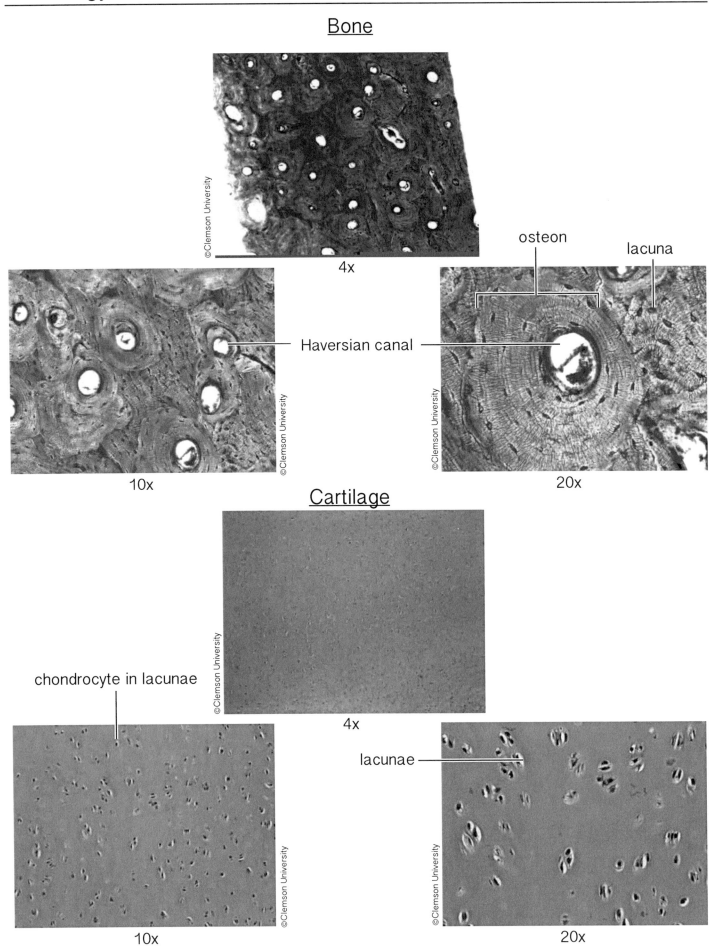

4x

10x

osteon

lacuna

Haversian canal

20x

Cartilage

4x

chondrocyte in lacunae

10x

lacunae

20x

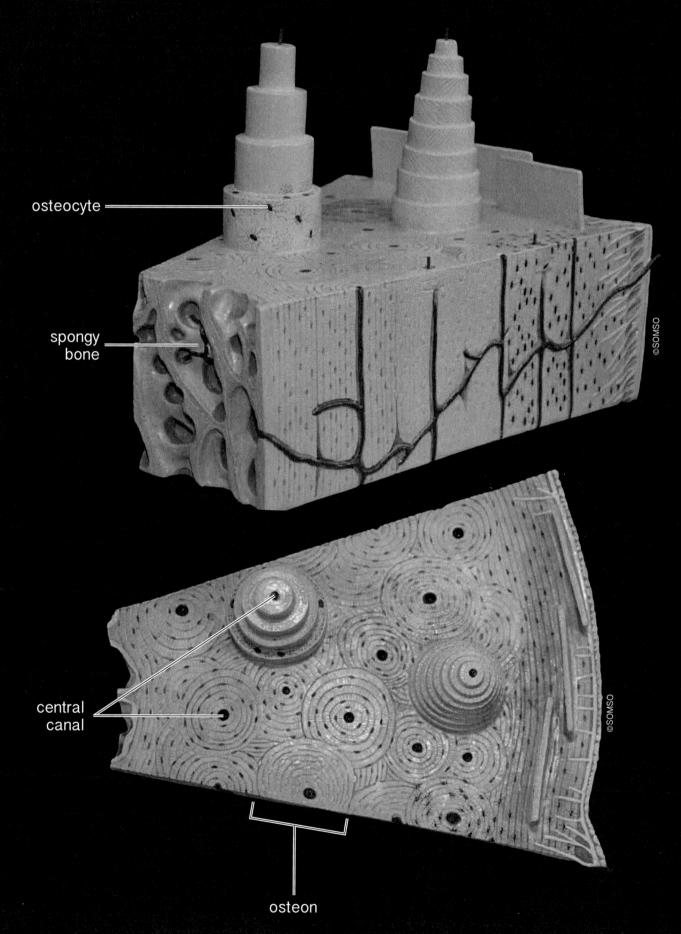

osteocyte

spongy
bone

©SOMSO

central
canal

©SOMSO

osteon

Canine - SKELETON OVERVIEW

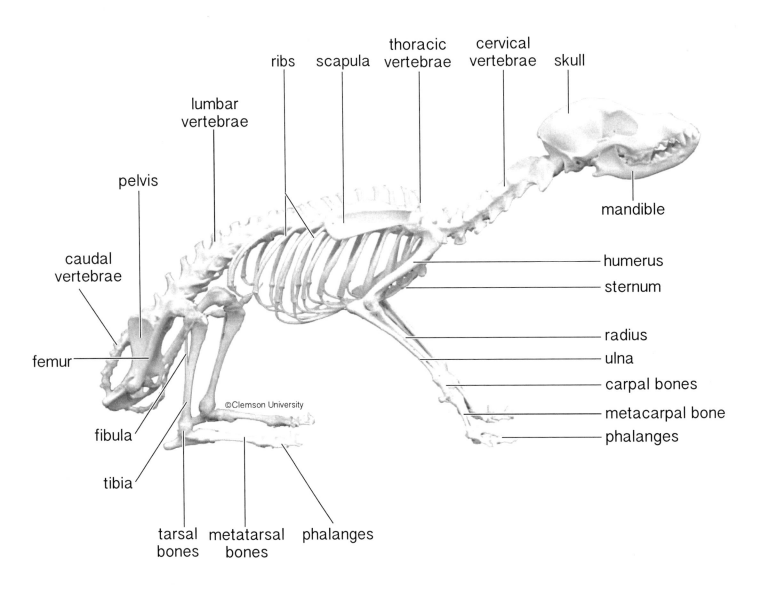

ribs scapula thoracic vertebrae cervical vertebrae skull

lumbar vertebrae

pelvis

mandible

caudal vertebrae

humerus

sternum

femur

radius

ulna

carpal bones

fibula

metacarpal bone

phalanges

tibia

tarsal bones   metatarsal bones   phalanges

©Clemson University

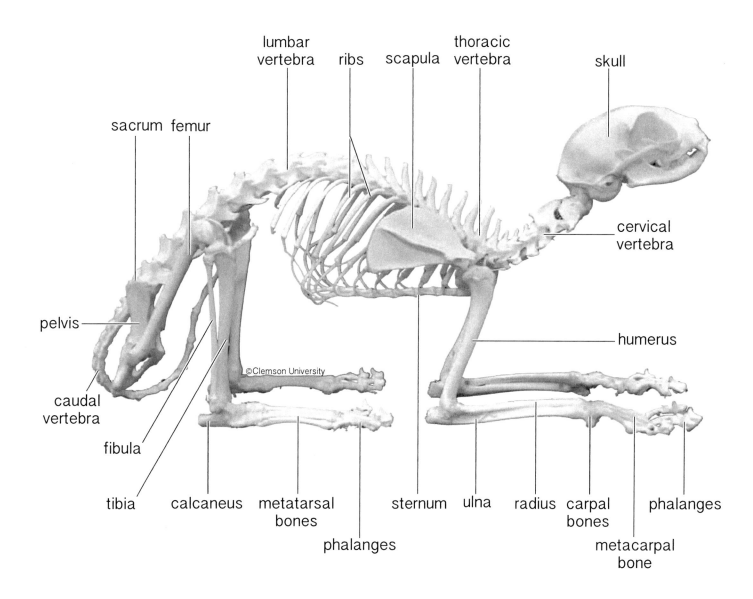

©Clemson University

Canine - SKULL

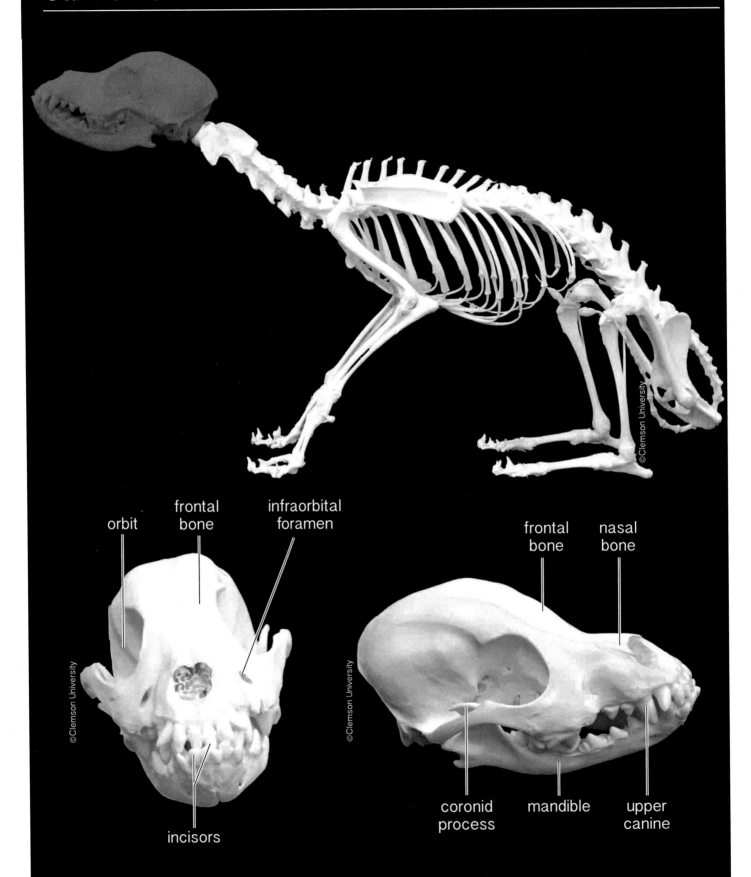

orbit

frontal bone

infraorbital foramen

©Clemson University

incisors

frontal bone

nasal bone

©Clemson University

coronid process

mandible

upper canine

©Clemson University

Bones of the Skull

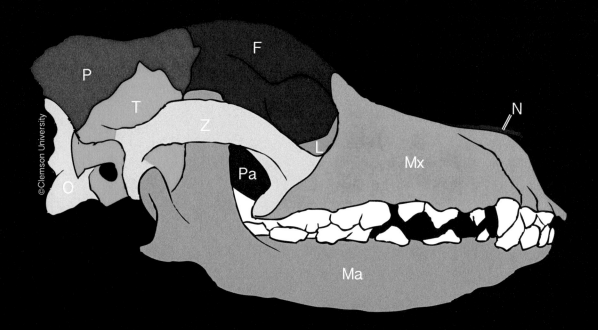

P - parietal
O - occipital
T - temporal
F - frontal
Z - zygomatic

L - lacrimal
Pa - palatine
Mx - maxilla
Ma - mandible
N - nasal

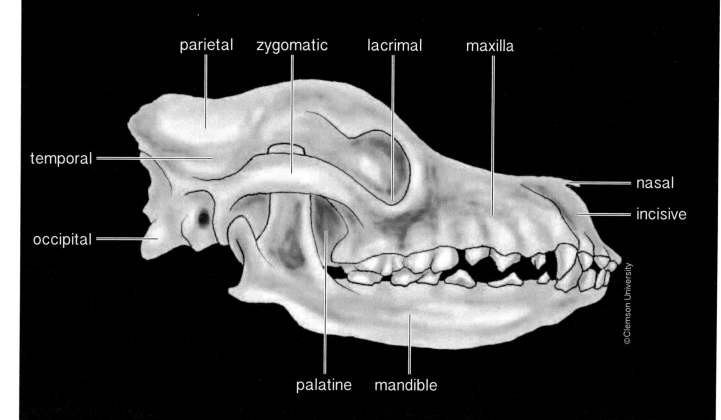

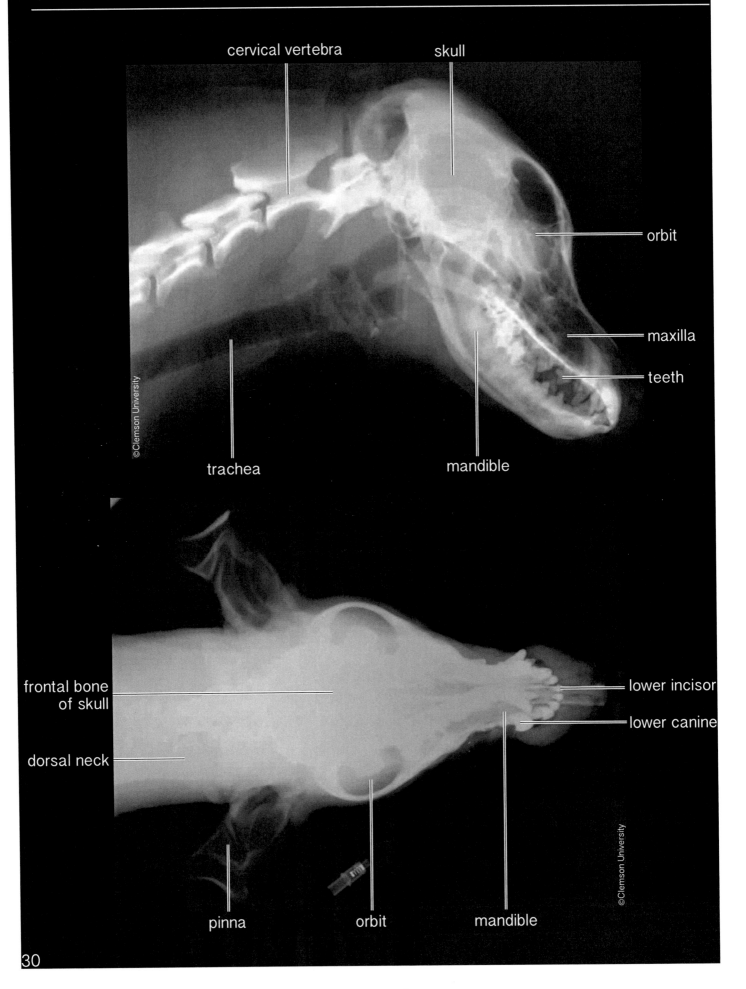

cervical vertebra

skull

orbit

maxilla

teeth

trachea

mandible

frontal bone
of skull

lower incisor

lower canine

dorsal neck

pinna

orbit

mandible

©Clemson University

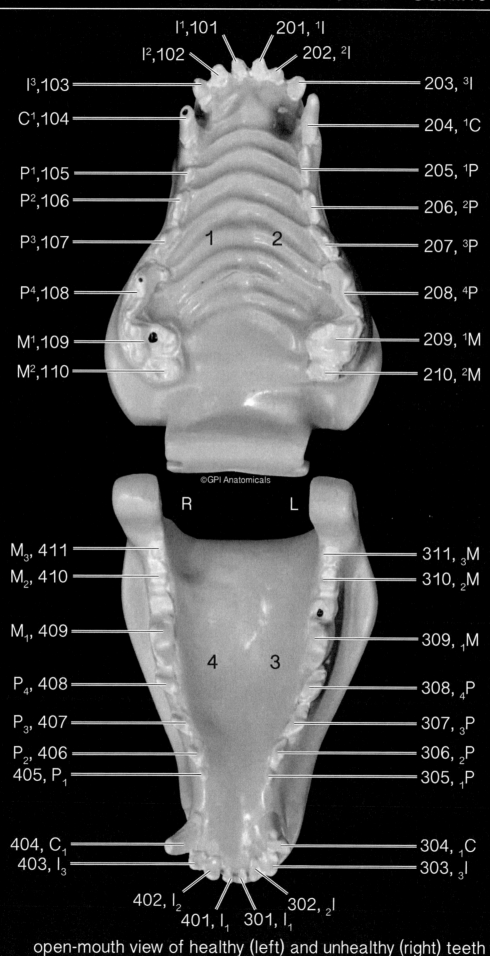

I¹,101 — $I^1, 101$
201, ¹I — $201, {}^1I$
I²,102 — $I^2, 102$
202, ²I — $202, {}^2I$
I³,103 — $I^3, 103$
203, ³I — $203, {}^3I$
C¹,104 — $C^1, 104$
204, ¹C — $204, {}^1C$
P¹,105 — $P^1, 105$
205, ¹P — $205, {}^1P$
P²,106 — $P^2, 106$
206, ²P — $206, {}^2P$
P³,107 — $P^3, 107$
207, ³P — $207, {}^3P$
P⁴,108 — $P^4, 108$
208, ⁴P — $208, {}^4P$
M¹,109 — $M^1, 109$
209, ¹M — $209, {}^1M$
M²,110 — $M^2, 110$
210, ²M — $210, {}^2M$

1   2

©GPI Anatomicals

R   L

I - incisor
C - canine
P - premolar
M - molar

M₃, 411 — $M_3, 411$
311, ₃M — $311, {}_3M$
M₂, 410 — $M_2, 410$
310, ₂M — $310, {}_2M$
M₁, 409 — $M_1, 409$
309, ₁M — $309, {}_1M$

4   3

P₄, 408 — $P_4, 408$
308, ₄P — $308, {}_4P$
P₃, 407 — $P_3, 407$
307, ₃P — $307, {}_3P$
P₂, 406 — $P_2, 406$
306, ₂P — $306, {}_2P$
405, P₁ — $405, P_1$
305, ₁P — $305, {}_1P$
404, C₁ — $404, C_1$
304, ₁C — $304, {}_1C$
403, I₃ — $403, I_3$
303, ₃I — $303, {}_3I$
402, I₂ — $402, I_2$
302, ₂I — $302, {}_2I$
401, I₁ — $401, I_1$
301, I₁ — $301, I_1$

open-mouth view of healthy (left) and unhealthy (right) teeth

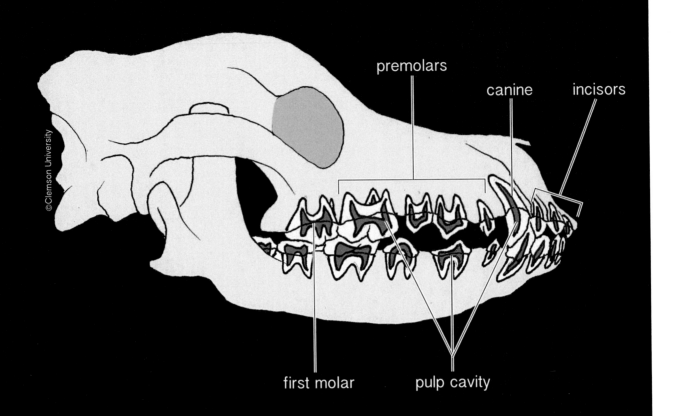

premolars

canine

incisors

first molar

pulp cavity

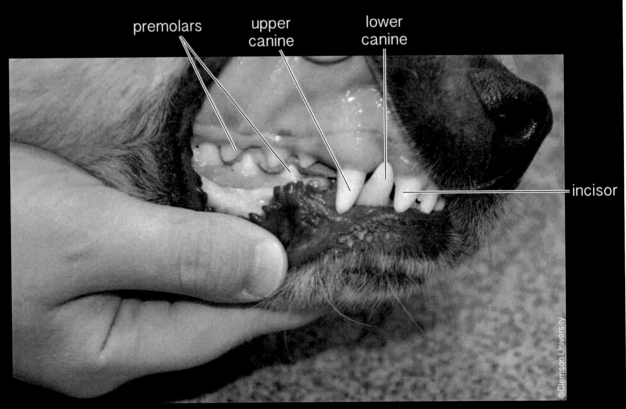

premolars

upper canine

lower canine

incisor

healthy teeth and gums

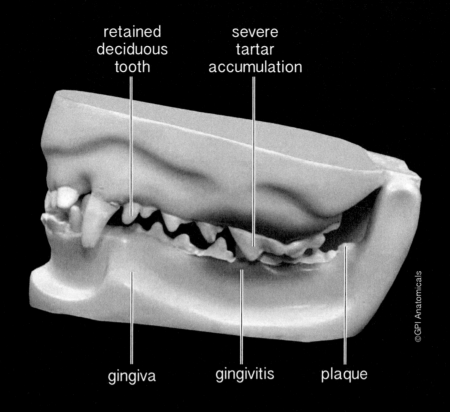

retained deciduous tooth

severe tartar accumulation

gingiva

gingivitis

plaque

©GPI Anatomicals

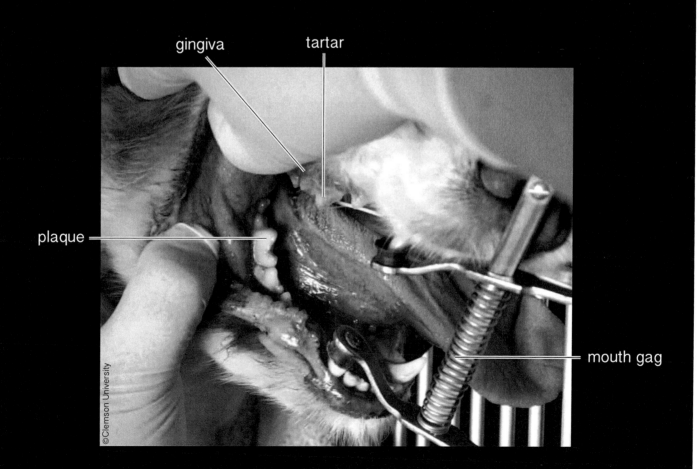

gingiva

tartar

plaque

mouth gag

©Clemson University

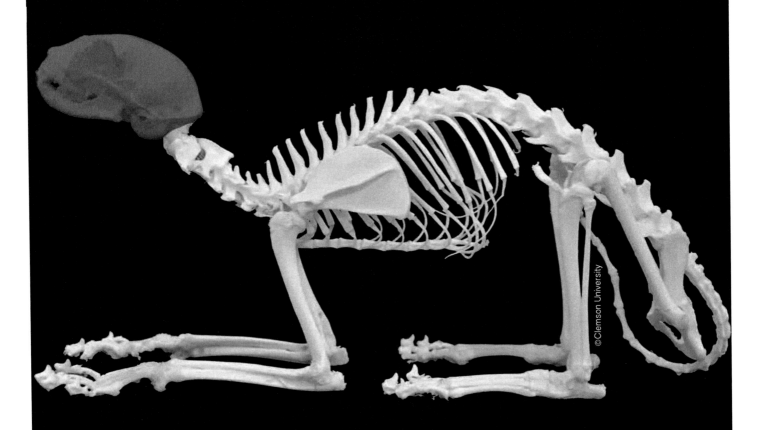

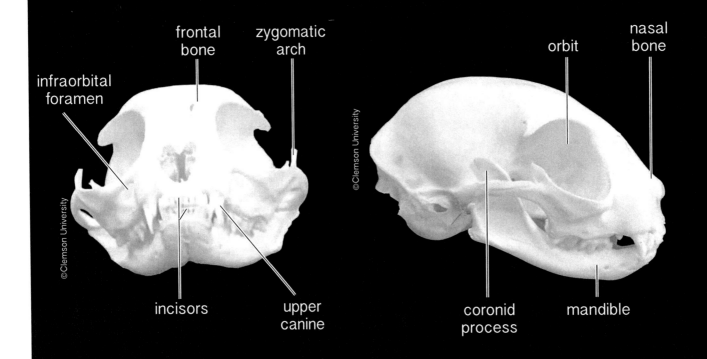

frontal bone

zygomatic arch

orbit

nasal bone

infraorbital foramen

incisors

upper canine

coronid process

mandible

Bones of the Skull

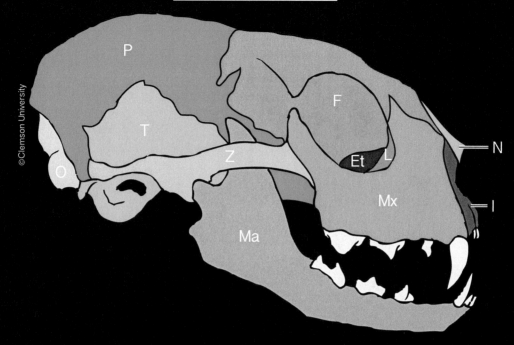

P - parietal   Z - zygomatic   Ma - mandible
O - occipital   L - lacrimal   N - nasal
T - temporal   Pa - palatine   Et - ethmoidal
F - frontal   Mx - maxilla   I-incisive

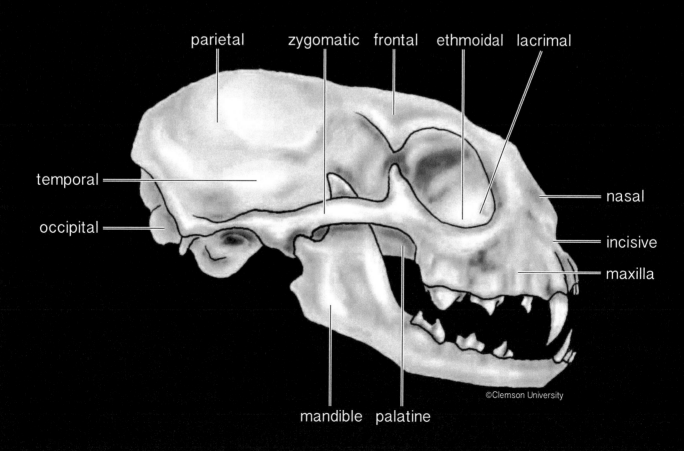

©Clemson University

35

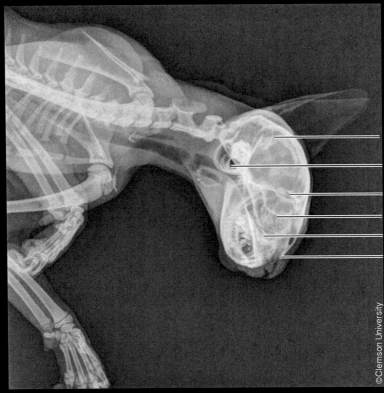

parietal bone
temporal bone
frontal bone
orbit
maxilla
nasal bone

lateral aspect

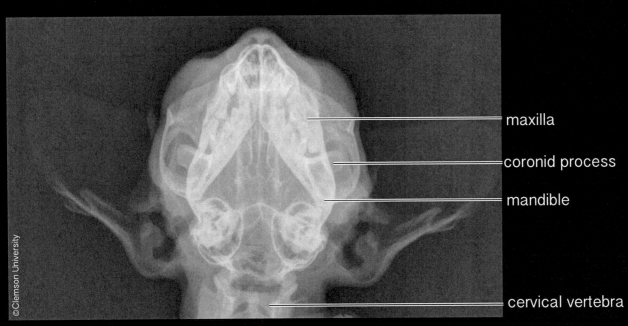

maxilla

coronid process

mandible

cervical vertebra

ventrodorsal aspect

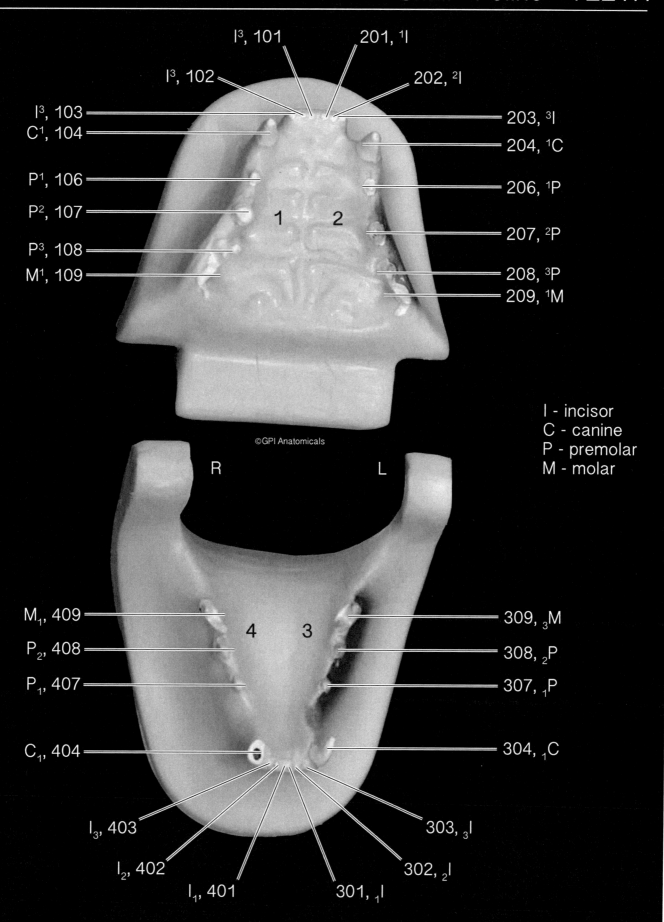

I³, 101
201, ¹I
I³, 102
202, ²I
I³, 103 — 203, ³I
C¹, 104 — 204, ¹C
P¹, 106 — 206, ¹P
P², 107 — 207, ²P
P³, 108 — 208, ³P
M¹, 109 — 209, ¹M

1   2

I - incisor
C - canine
P - premolar
M - molar

©GPI Anatomicals

R   L

M_1, 409 — 309, $_3M$
P_2, 408 — 308, $_2P$
P_1, 407 — 307, $_1P$

4   3

C_1, 404 — 304, $_1C$

I_3, 403
303, $_3I$
I_2, 402
302, $_2I$
I_1, 401
301, $_1I$

open-mouth view of healthy (left) and unhealthy (right) teeth

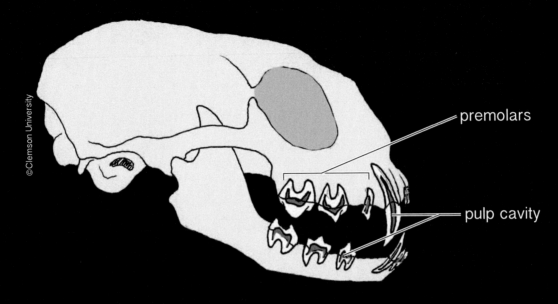

premolars

pulp cavity

©Clemson University

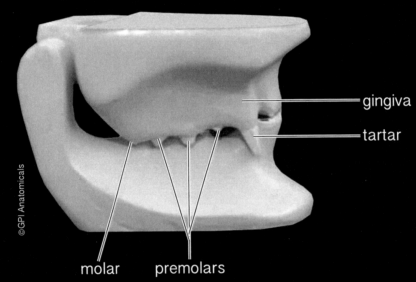

gingiva

tartar

molar   premolars

©GPI Anatomicals

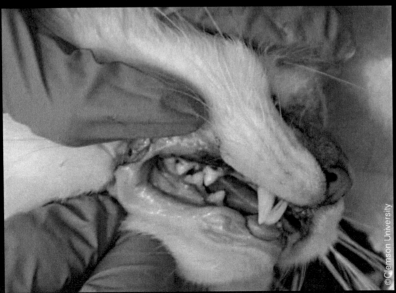

©Clemson University

lateral view of permanent teeth with recessed
gums and exposed tooth roots

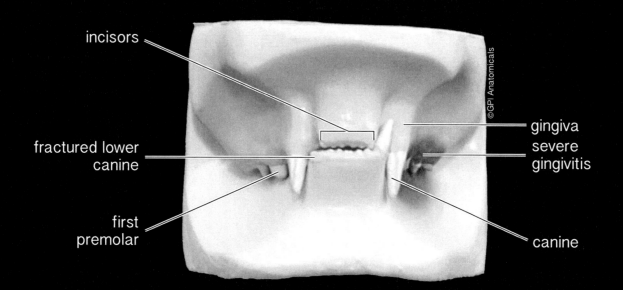

incisors

fractured lower
canine

first
premolar

gingiva
severe
gingivitis

canine

©GPI Anatomicals

frontal view of permanent (adult) teeth

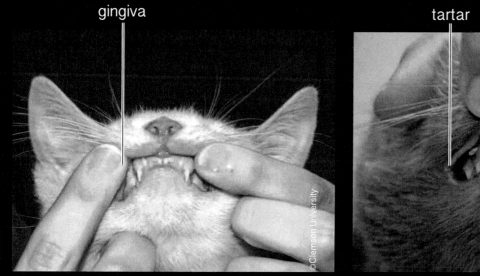

gingiva

tartar

frontal view of deciduous teeth

frontal view of permanent teeth

©Clemson University

©Clemson University

VERTEBRAL COLUMN

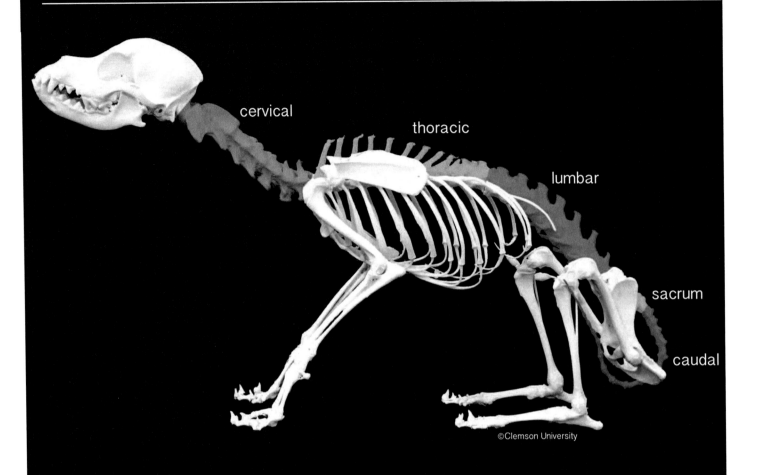

cervical

thoracic

lumbar

sacrum

caudal

©Clemson University

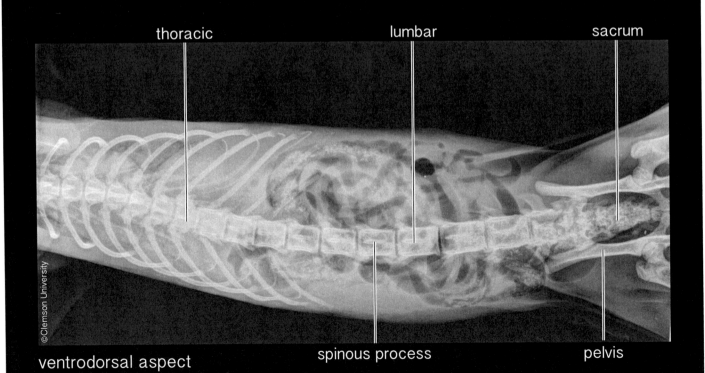

thoracic

lumbar

sacrum

©Clemson University

ventrodorsal aspect

spinous process

pelvis

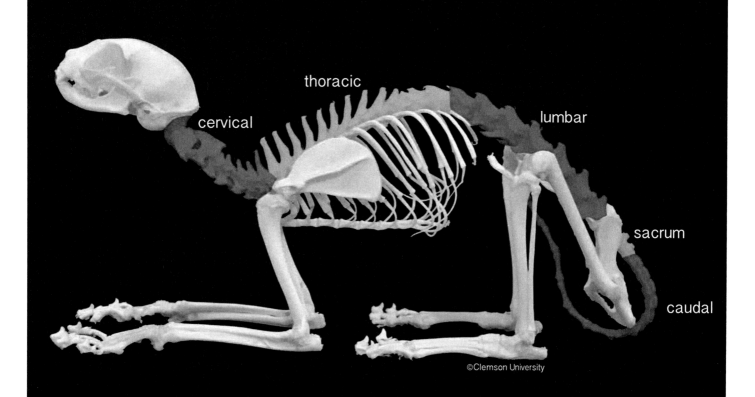

thoracic

cervical

lumbar

sacrum

caudal

©Clemson University

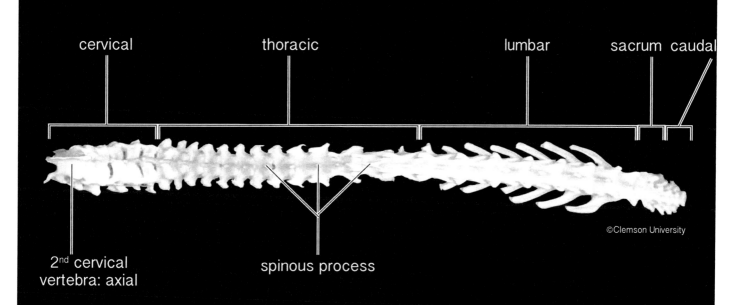

cervical

thoracic

lumbar

sacrum caudal

2nd cervical
vertebra: axial

spinous process

©Clemson University

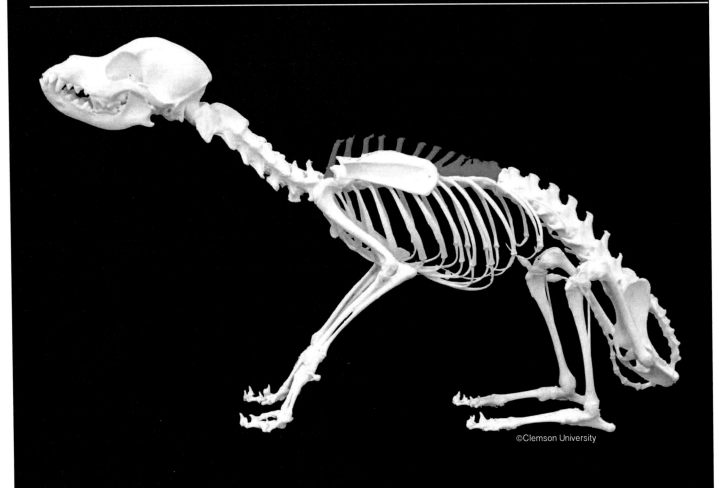

©Clemson University

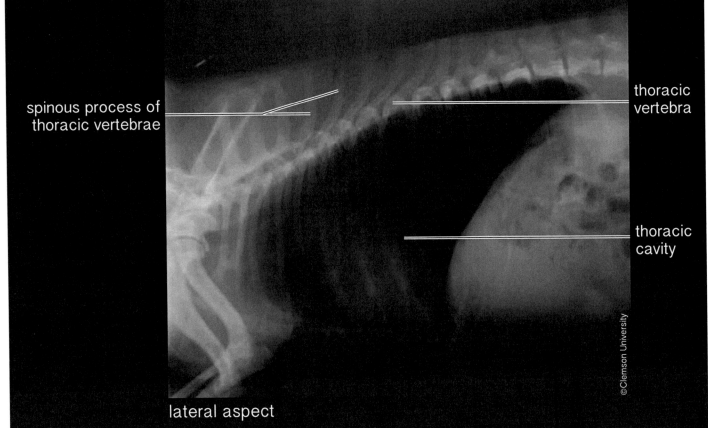

spinous process of
thoracic vertebrae

thoracic
vertebra

thoracic
cavity

©Clemson University

lateral aspect

©Clemson University

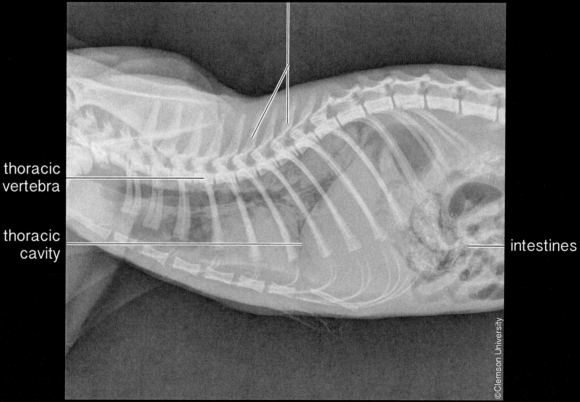

spinous process of
thoracic vertebrae

thoracic
vertebra

thoracic
cavity

intestines

©Clemson University

left lateral aspect

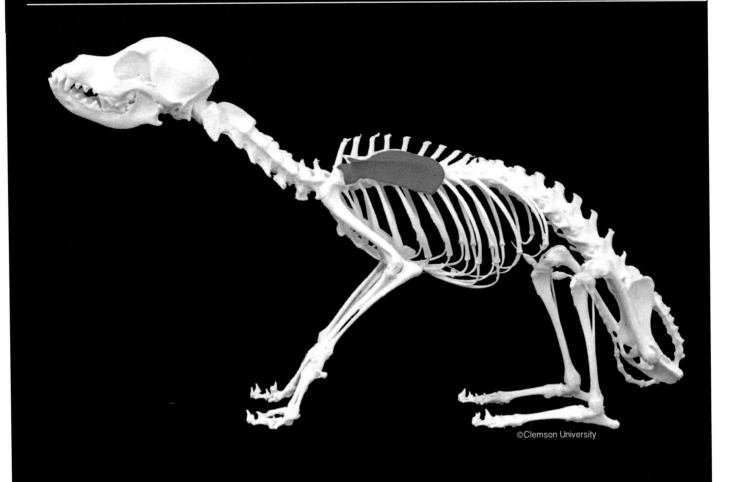

©Clemson University

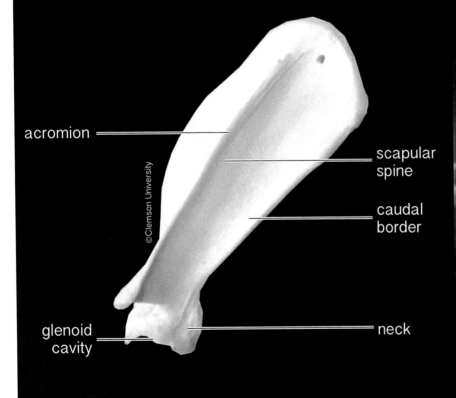

acromion

scapular spine

caudal border

neck

glenoid cavity

©Clemson University

lateral aspect of left scapula

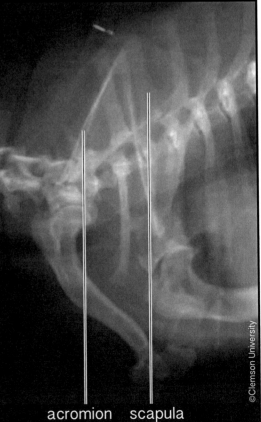

acromion   scapula

©Clemson University

©Clemson University

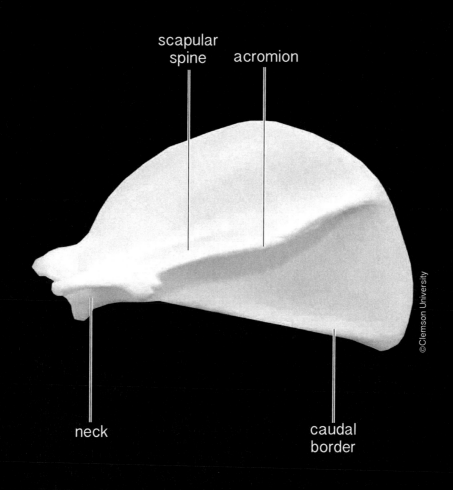

scapular spine

acromion

neck

caudal border

©Clemson University

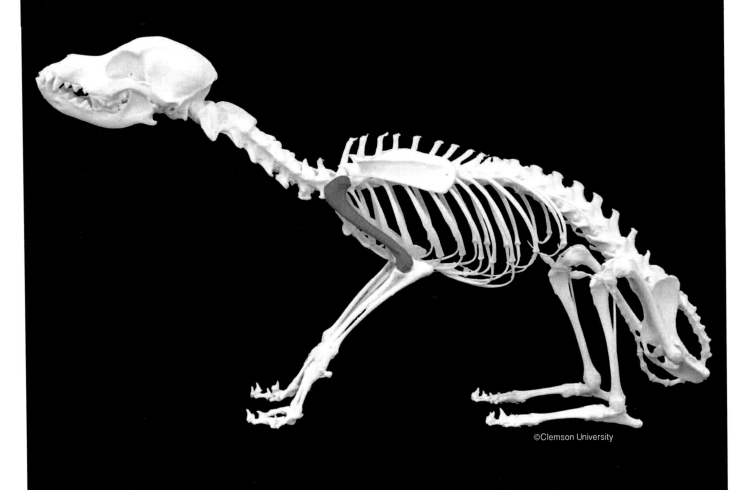

©Clemson University

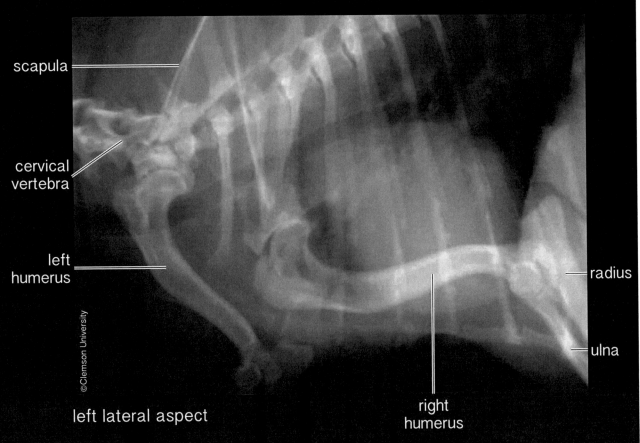

scapula

cervical
vertebra

left
humerus

radius

ulna

©Clemson University

left lateral aspect

right
humerus

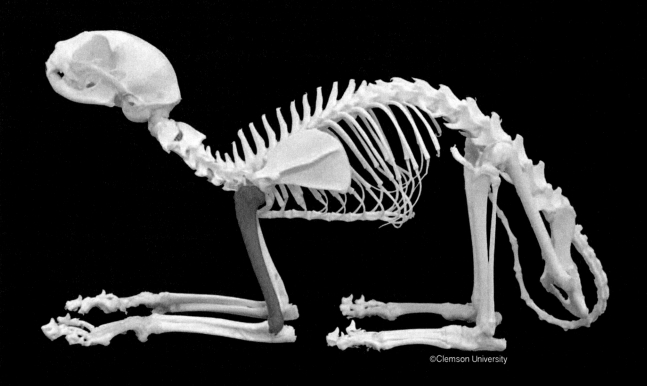

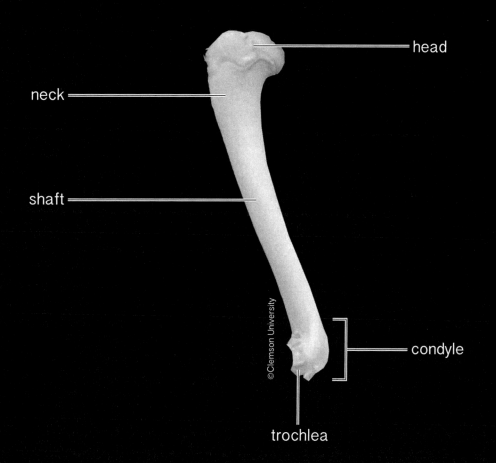

head

neck

shaft

condyle

trochlea

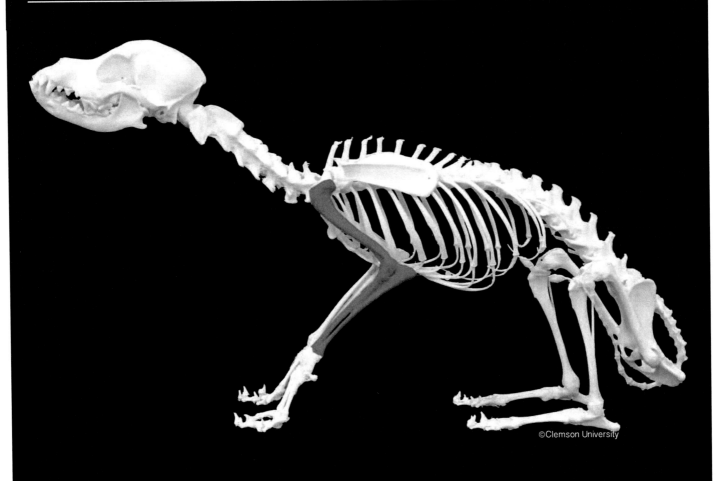

©Clemson University

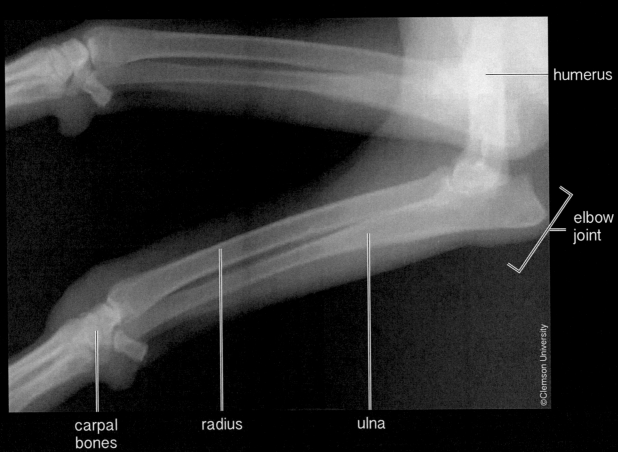

humerus

elbow joint

carpal bones

radius

ulna

©Clemson University

©Clemson University

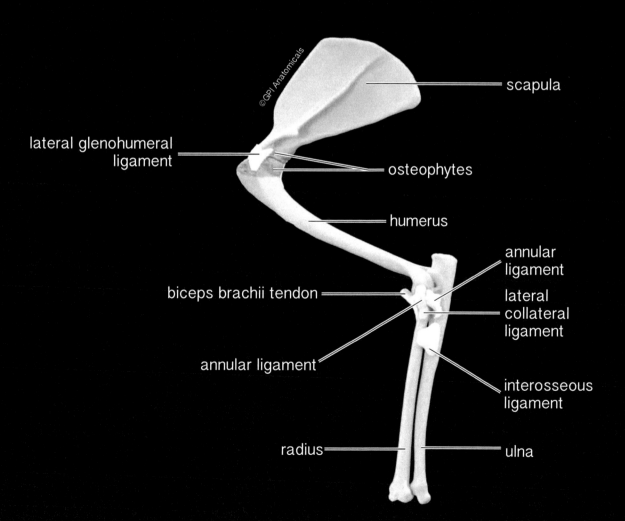

©GPI Anatomicals

scapula

lateral glenohumeral ligament

osteophytes

humerus

annular ligament

biceps brachii tendon

lateral collateral ligament

annular ligament

interosseous ligament

radius

ulna

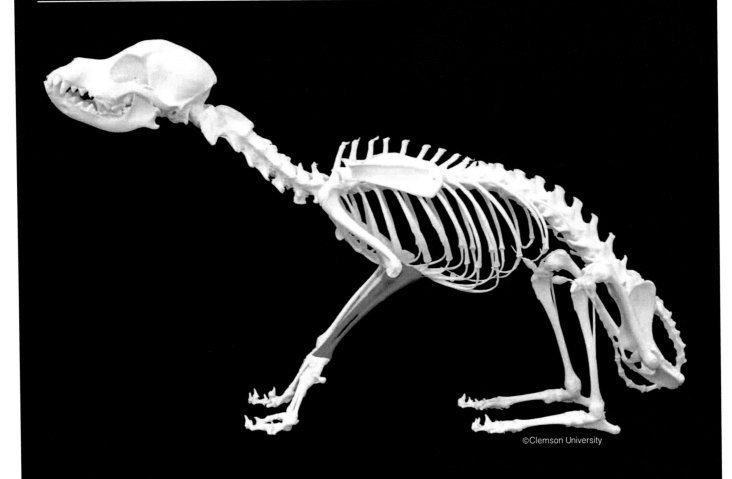

©Clemson University

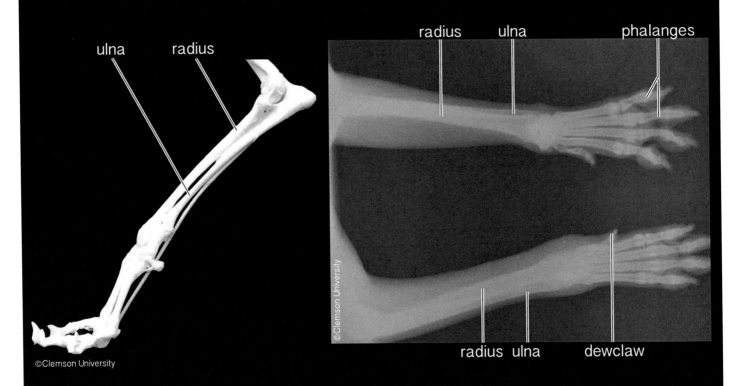

ulna radius

radius ulna phalanges

©Clemson University

radius ulna dewclaw

©Clemson University

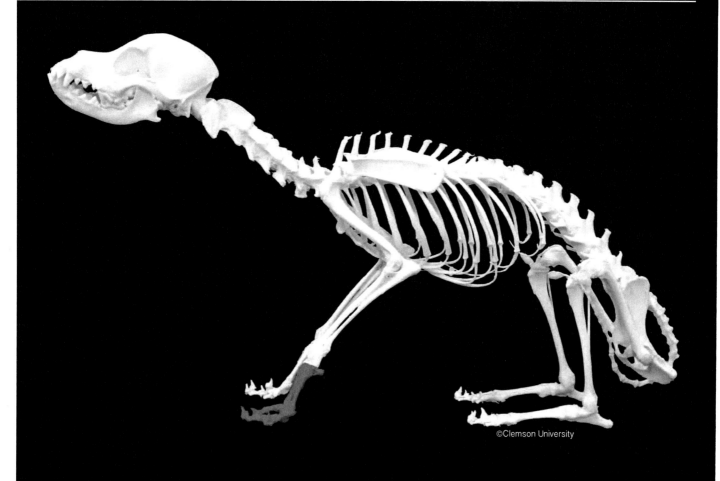

©Clemson University

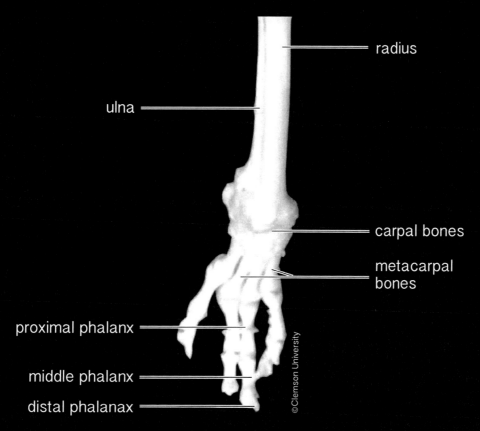

radius

ulna

carpal bones

metacarpal bones

proximal phalanx

middle phalanx

distal phalanax

©Clemson University

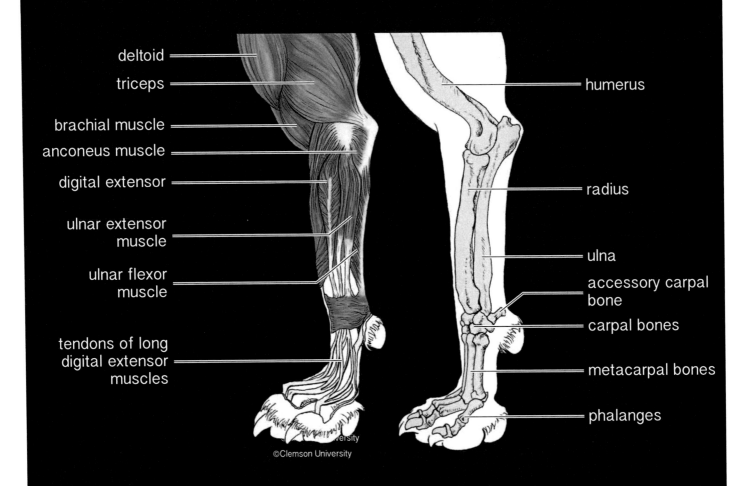

deltoid

triceps

brachial muscle

anconeus muscle

digital extensor

ulnar extensor muscle

ulnar flexor muscle

tendons of long digital extensor muscles

humerus

radius

ulna

accessory carpal bone

carpal bones

metacarpal bones

phalanges

©Clemson University

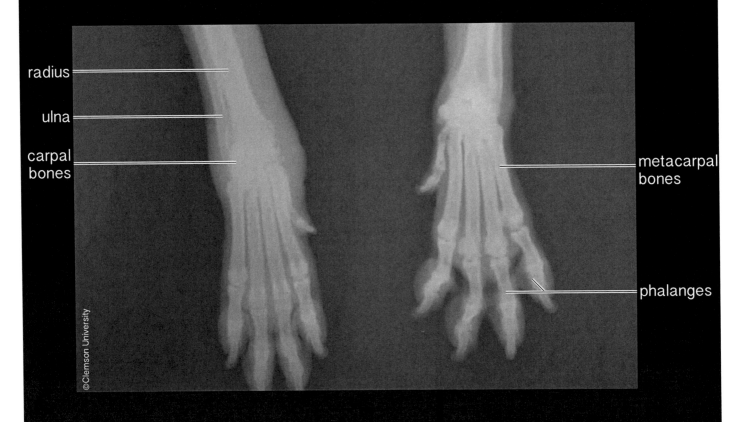

radius

ulna

carpal bones

metacarpal bones

phalanges

©Clemson University

©Clemson University

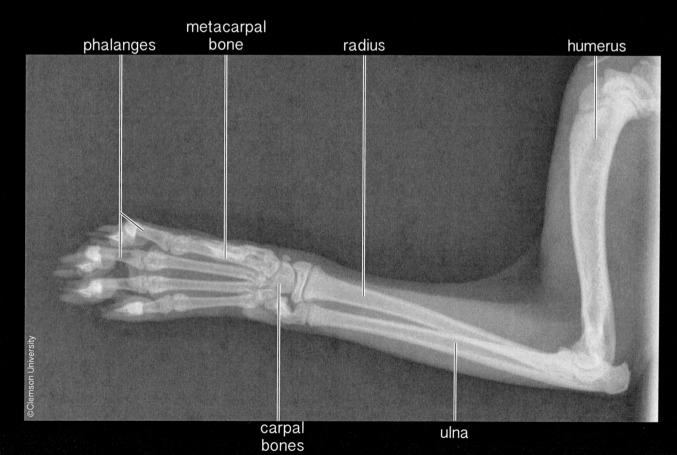

phalanges

metacarpal
bone

radius

humerus

carpal
bones

ulna

©Clemson University

©Clemson University

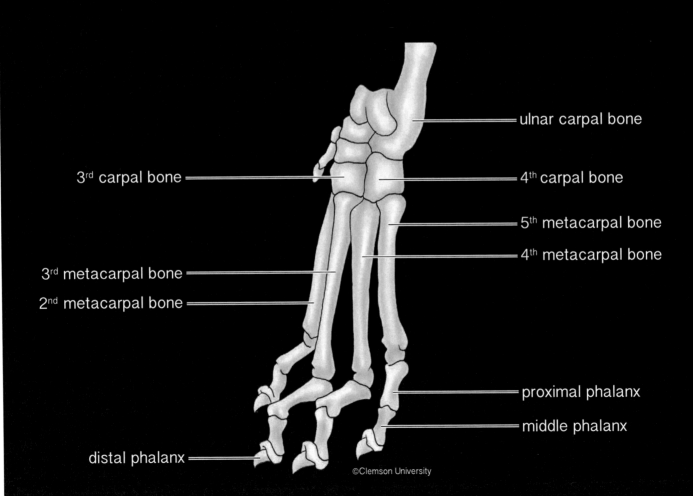

ulnar carpal bone

3rd carpal bone

4th carpal bone

5th metacarpal bone

4th metacarpal bone

3rd metacarpal bone

2nd metacarpal bone

proximal phalanx

middle phalanx

distal phalanx

©Clemson University

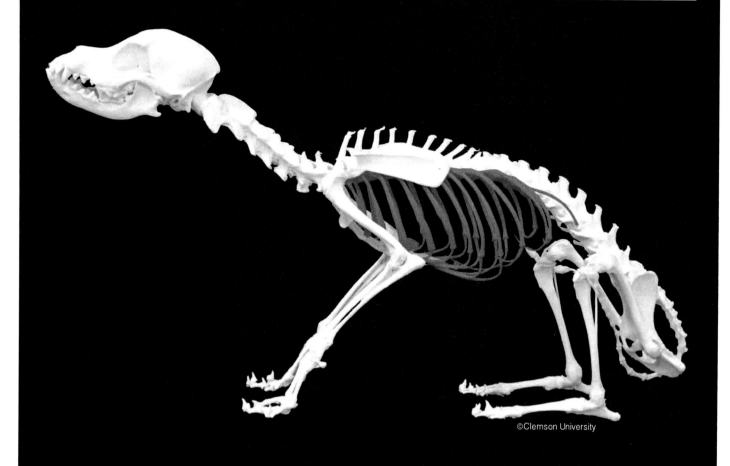

thoracic vertebra

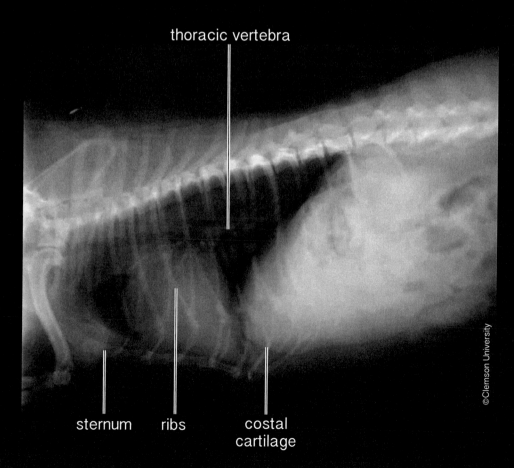

sternum ribs costal cartilage

©Clemson University

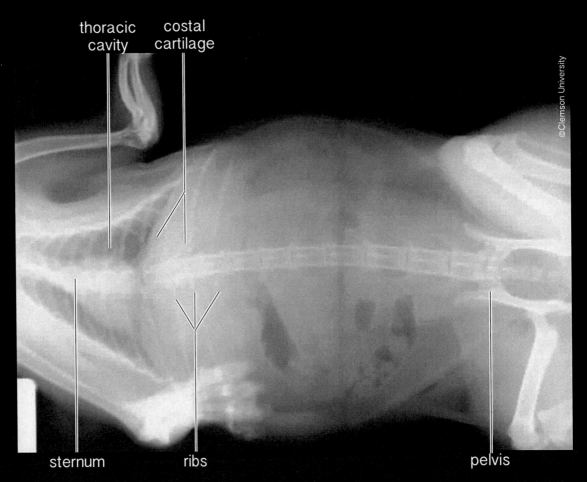

©Clemson University

thoracic
cavity

costal
cartilage

sternum

ribs

pelvis

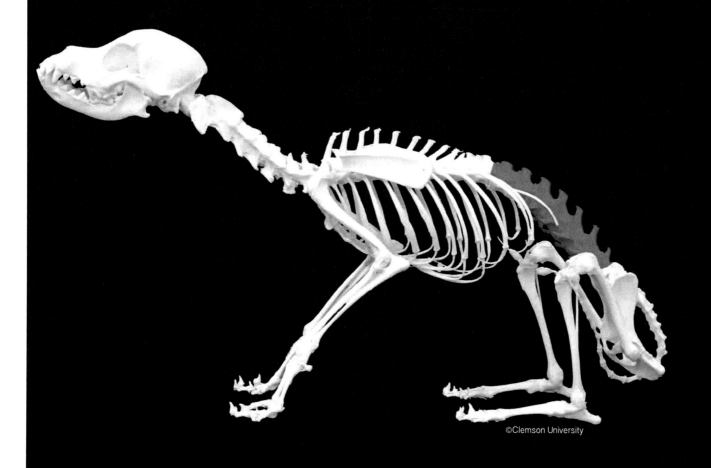

©Clemson University

©Clemson University

Vertebral Column - LUMBAR VERTEBRAE

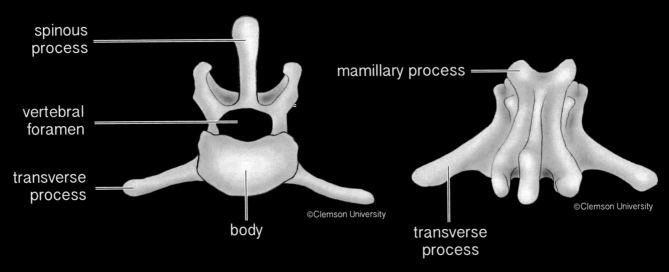

spinous process

vertebral foramen

transverse process

body

mamillary process

transverse process

©Clemson University

©Clemson University

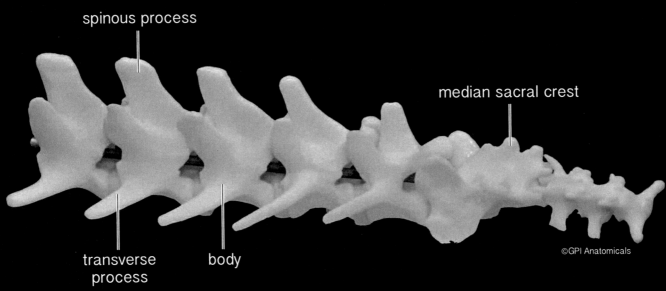

spinous process

median sacral crest

transverse process

body

©GPI Anatomicals

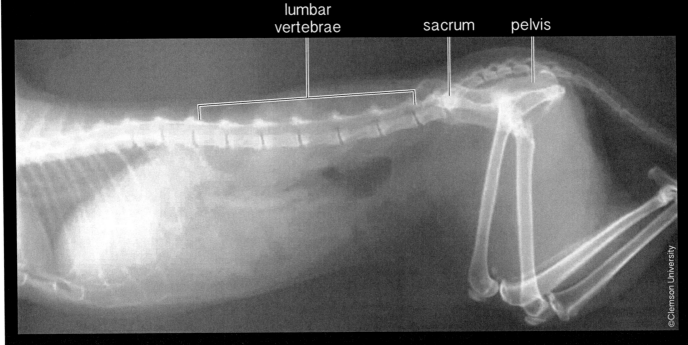

lumbar vertebrae

sacrum

pelvis

©Clemson University

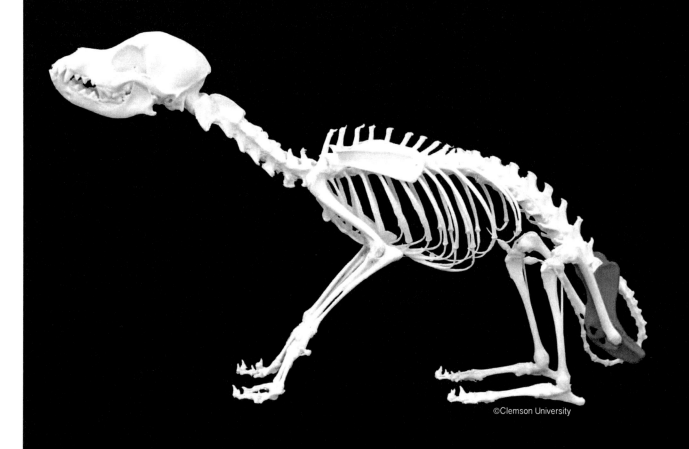

©Clemson University

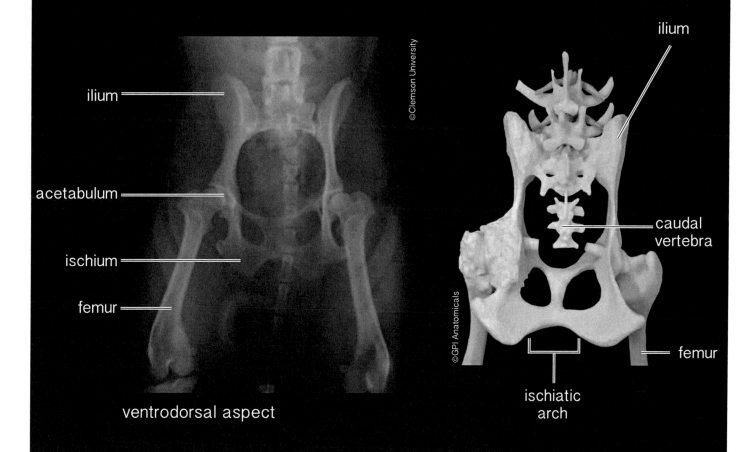

©Clemson University

ilium

acetabulum

ischium

femur

ventrodorsal aspect

ilium

caudal
vertebra

©GPI Anatomicals

femur

ischiatic
arch

©Clemson University

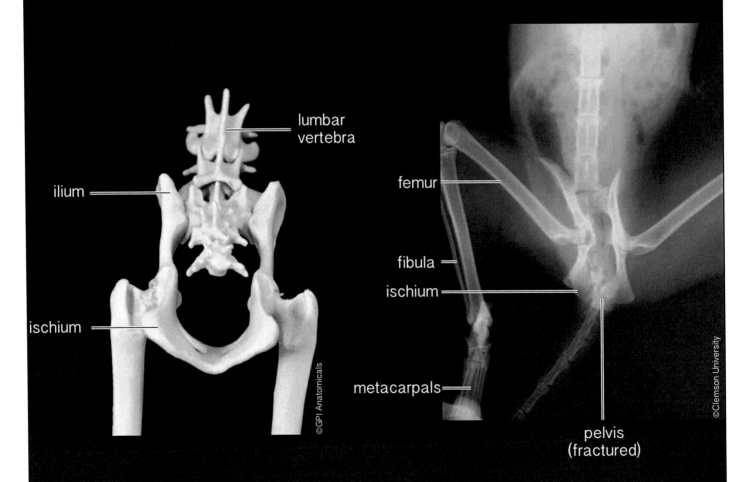

lumbar vertebra

ilium

ischium

femur

fibula

ischium

metacarpals

pelvis (fractured)

©GPI Anatomicals

©Clemson University

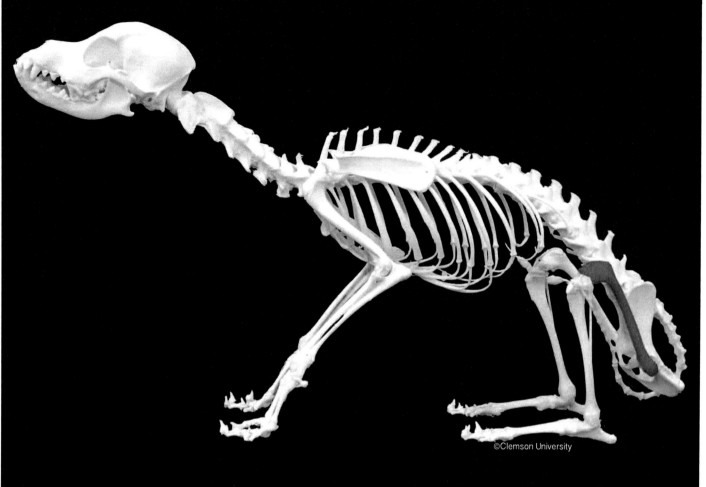

©Clemson University

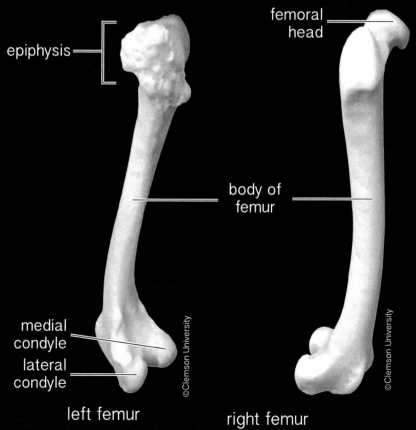

epiphysis

femoral head

body of femur

medial condyle

lateral condyle

©Clemson University

left femur

right femur

©Clemson University

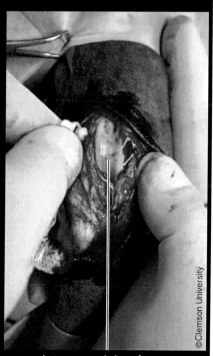

©Clemson University

intercondylar fossa
(distal end of femur)

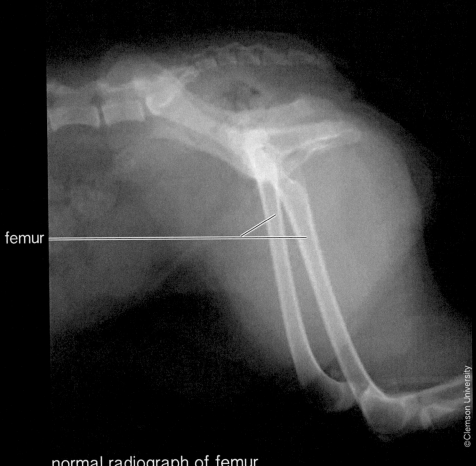

femur

normal radiograph of femur

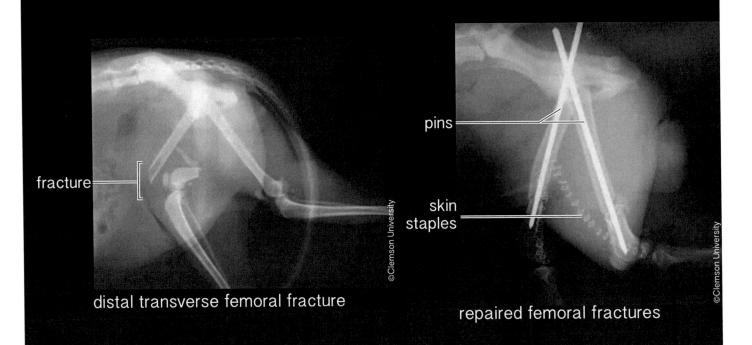

fracture

distal transverse femoral fracture

pins

skin
staples

repaired femoral fractures

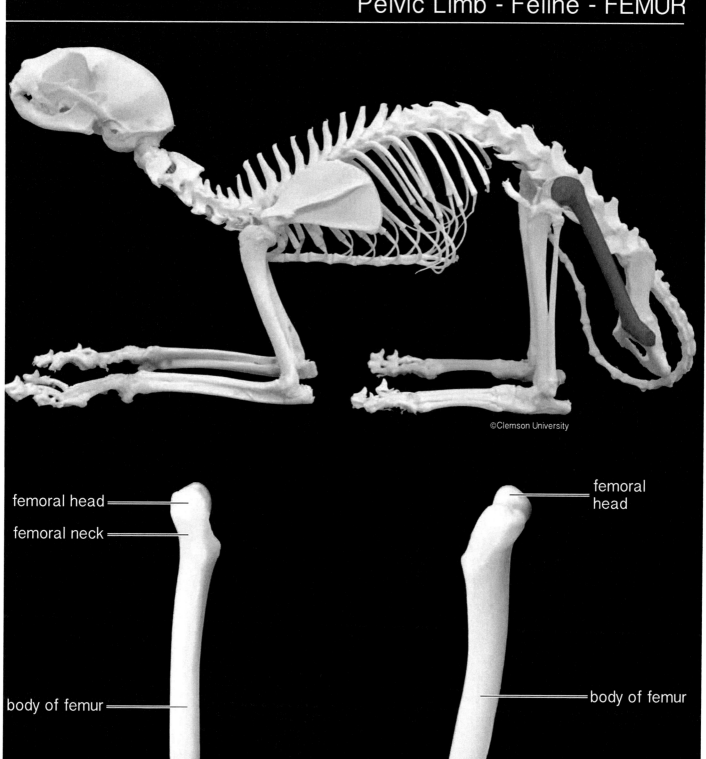

©Clemson University

femoral head

femoral neck

femoral head

body of femur

body of femur

medial condyle

lateral condyle

©Clemson University

©Clemson University

left femur, lateral aspect

right femur, lateral aspect

Pelvic Limb - STIFLE JOINT

Canine

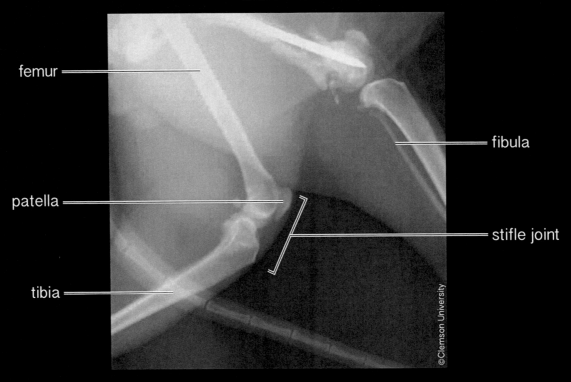

femur

fibula

patella

stifle joint

tibia

©Clemson University

Feline

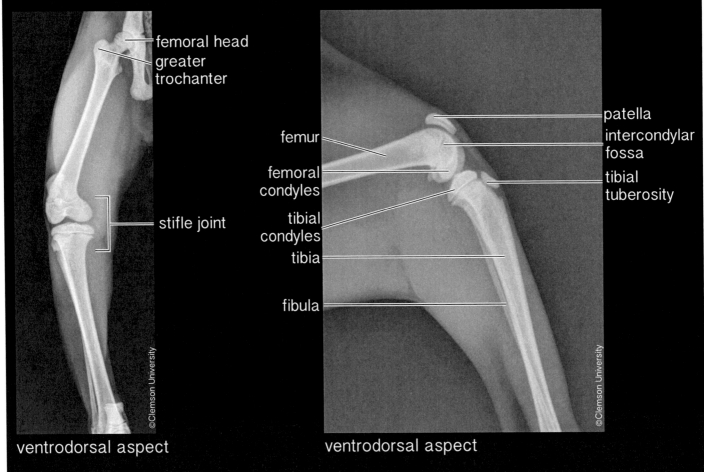

femoral head
greater
trochanter

patella
intercondylar
fossa

femur

femoral
condyles

tibial
tuberosity

tibial
condyles

stifle joint

tibia

fibula

©Clemson University

©Clemson University

ventrodorsal aspect

ventrodorsal aspect

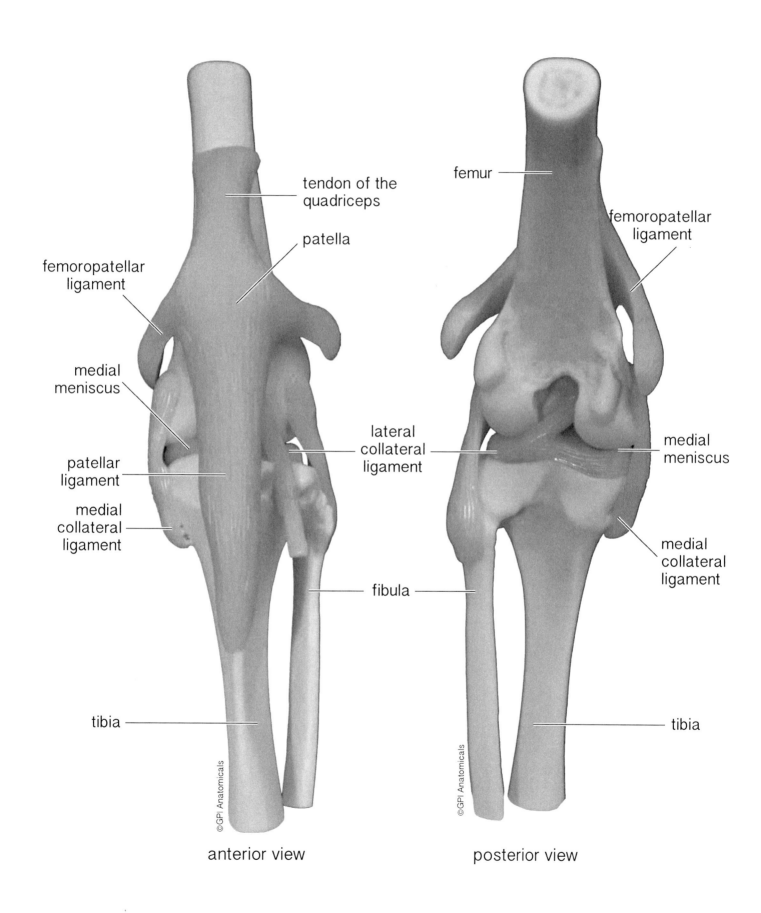

tendon of the quadriceps

patella

femoropatellar ligament

medial meniscus

patellar ligament

medial collateral ligament

tibia

femur

femoropatellar ligament

lateral collateral ligament

medial meniscus

medial collateral ligament

fibula

tibia

©GPI Anatomicals

©GPI Anatomicals

anterior view

posterior view

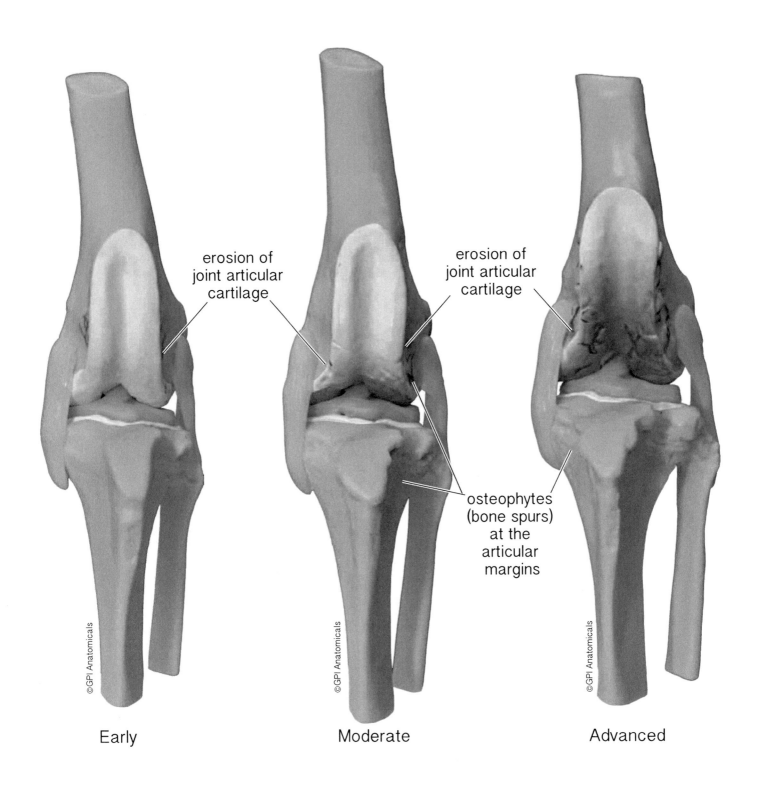

erosion of
joint articular
cartilage

erosion of
joint articular
cartilage

osteophytes
(bone spurs)
at the
articular
margins

©GPI Anatomicals

©GPI Anatomicals

©GPI Anatomicals

Early

Moderate

Advanced

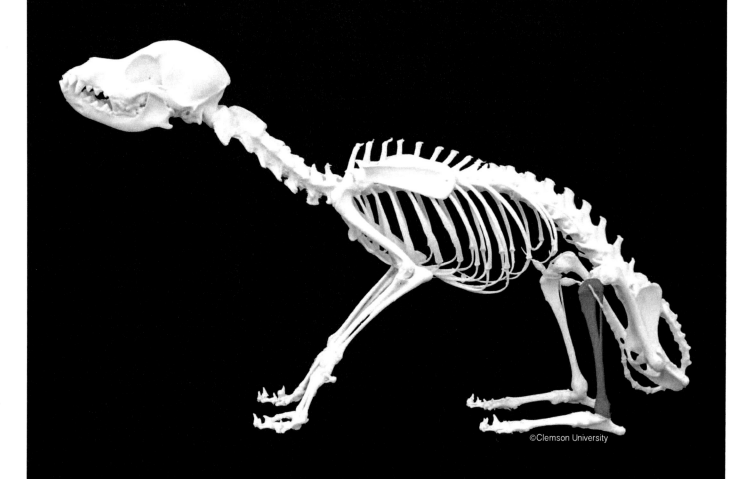

©Clemson University

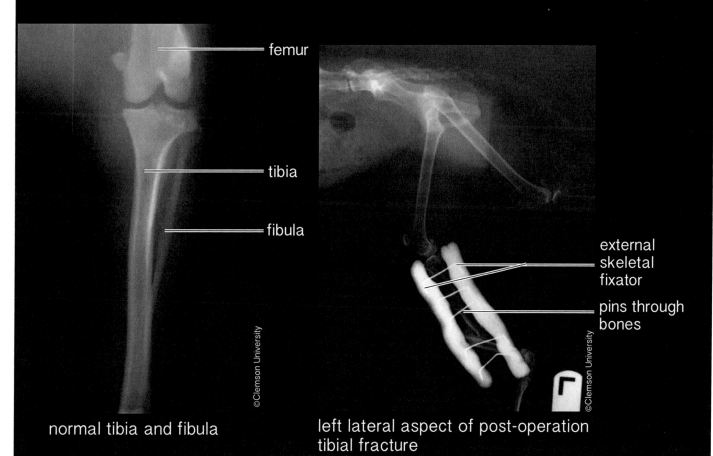

femur

tibia

fibula

external skeletal fixator

pins through bones

©Clemson University

©Clemson University

normal tibia and fibula

left lateral aspect of post-operation tibial fracture

©Clemson University

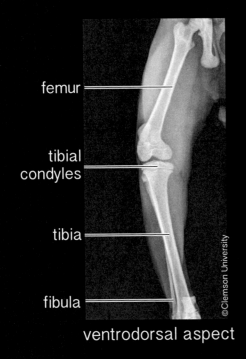

femur

tibial
condyles

tibia

fibula

ventrodorsal aspect

©Clemson University

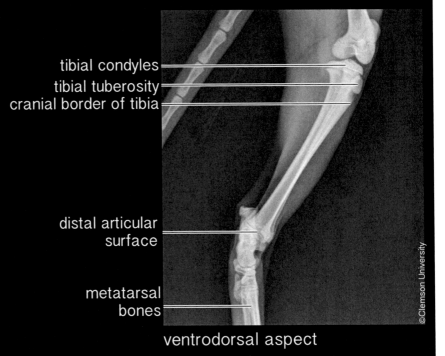

tibial condyles

tibial tuberosity

cranial border of tibia

distal articular
surface

metatarsal
bones

ventrodorsal aspect

©Clemson University

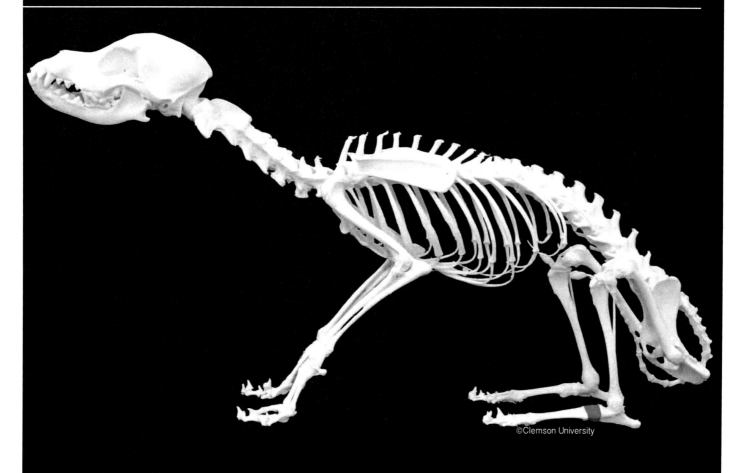

©Clemson University

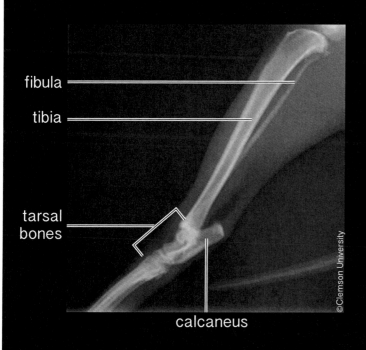

fibula

tibia

tarsal
bones

calcaneus

©Clemson University

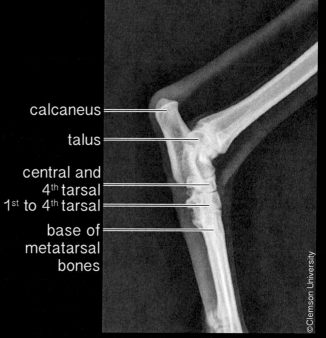

calcaneus

talus

central and
4th tarsal
1st to 4th tarsal

base of
metatarsal
bones

©Clemson University

left lateral aspect

©Clemson University

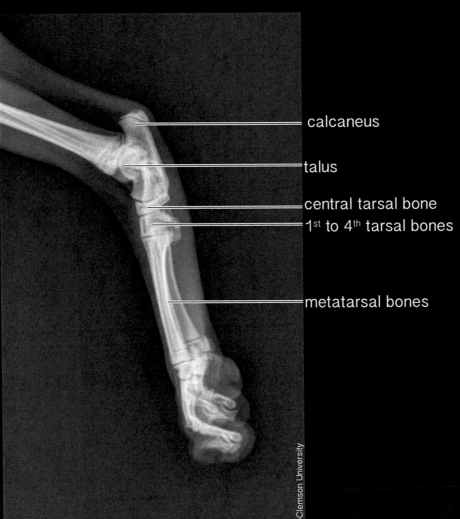

calcaneus

talus

central tarsal bone

1st to 4th tarsal bones

metatarsal bones

©Clemson University

right lateral aspect

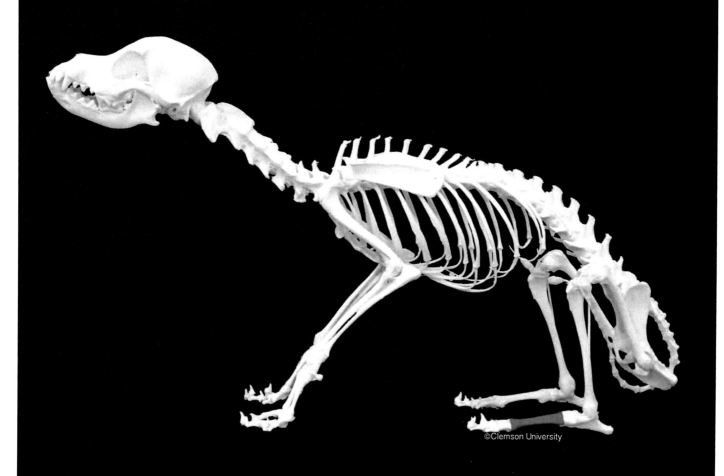

©Clemson University

©Clemson University

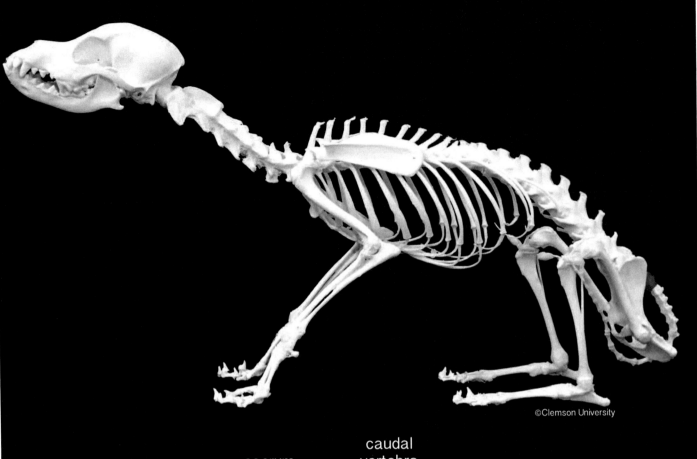

©Clemson University

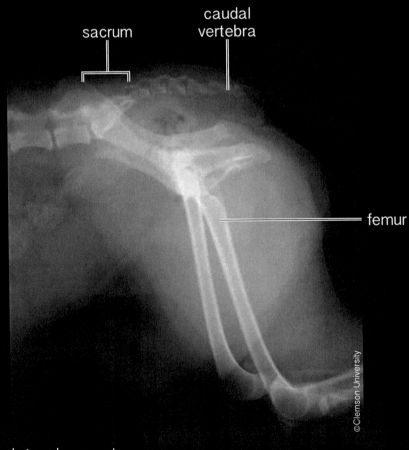

sacrum

caudal
vertebra

femur

©Clemson University

lateral aspect

©Clemson University

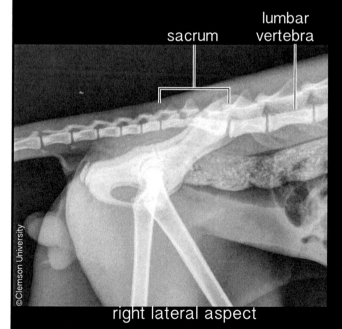

sacrum

lumbar
vertebra

©Clemson University

right lateral aspect

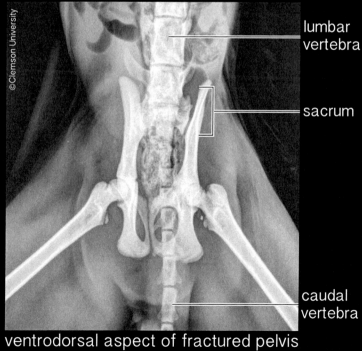

©Clemson University

lumbar
vertebra

sacrum

caudal
vertebra

ventrodorsal aspect of fractured pelvis

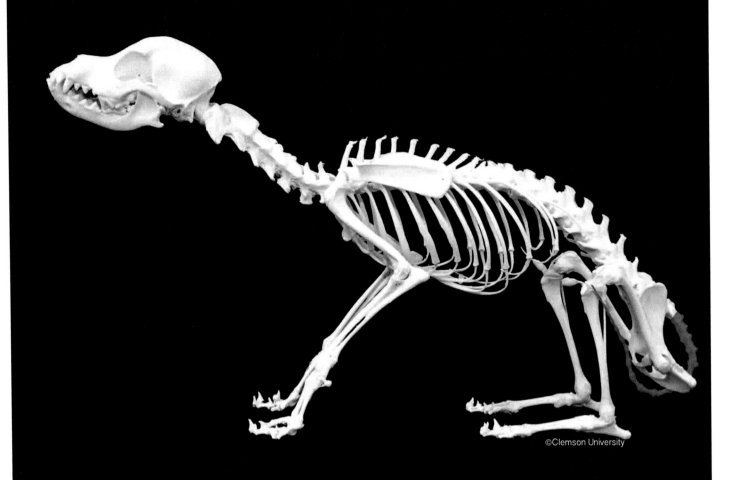

©Clemson University

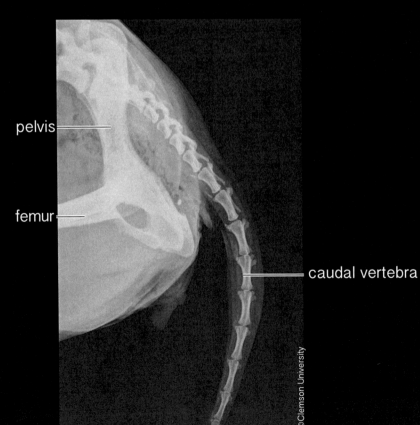

pelvis

femur

caudal vertebra

©Clemson University

left lateral aspect

©Clemson University

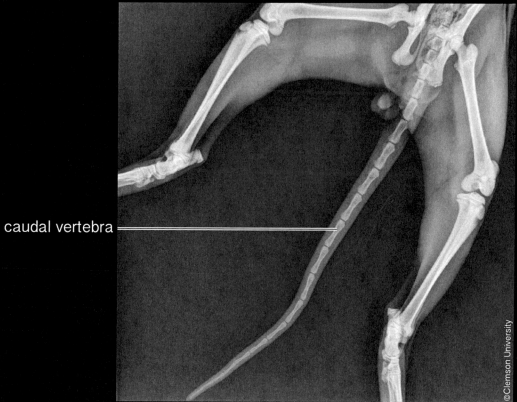

caudal vertebra

©Clemson University

ventrodorsal aspect

Chapter 4

Muscular System

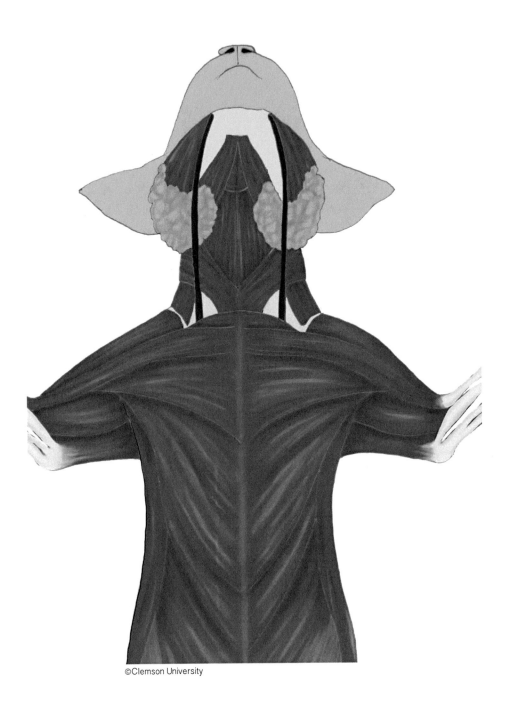

©Clemson University

Smooth Muscle

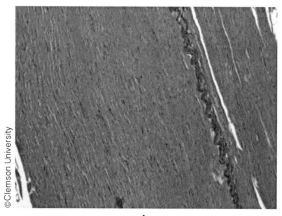

4x

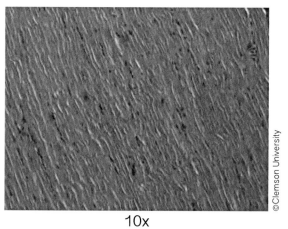

10x

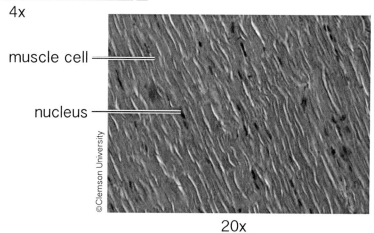

muscle cell

nucleus

20x

Skeletal Muscle

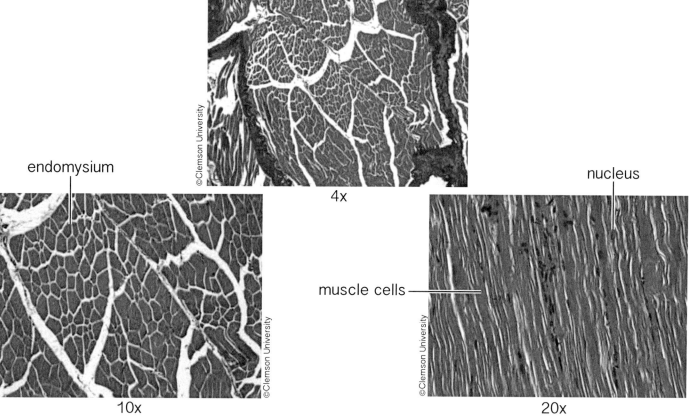

4x

endomysium

10x

nucleus

muscle cells

20x

SARCOMERE OF SKELETAL MUSCLE CELL

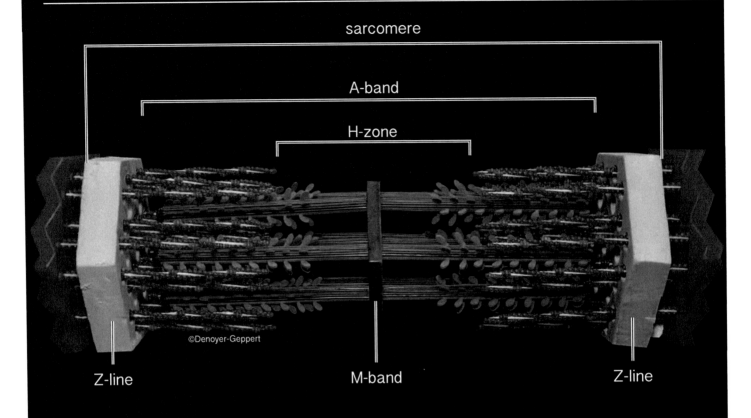

sarcomere

A-band

H-zone

Z-line

M-band

Z-line

©Denoyer-Geppert

thin filament   thick filament

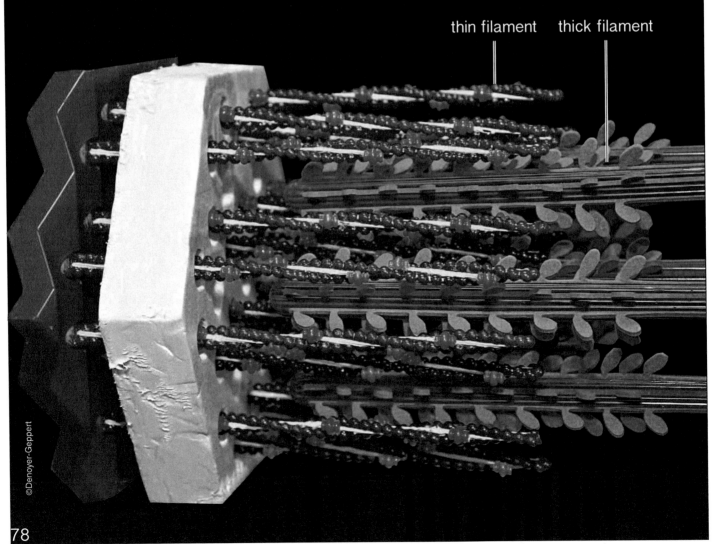

©Denoyer-Geppert

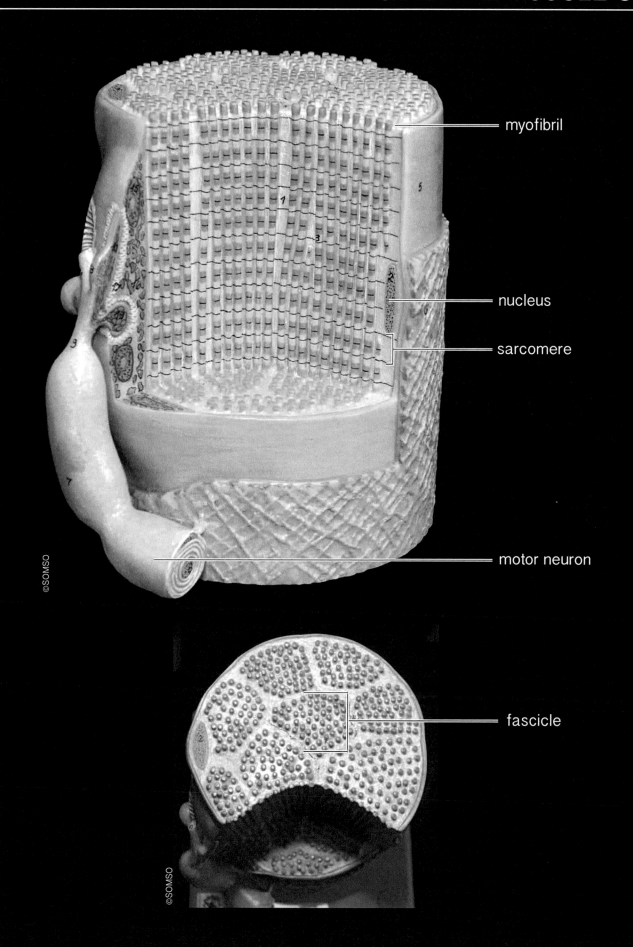

myofibril

nucleus

sarcomere

motor neuron

fascicle

©SOMSO

©SOMSO

SKELETAL MUSCLE FIBER

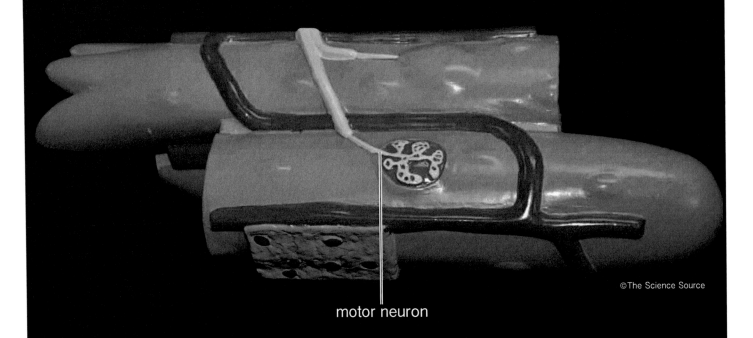

motor neuron

©The Science Source

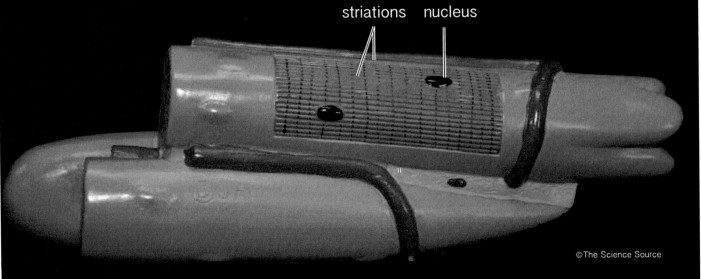

striations   nucleus

©The Science Source

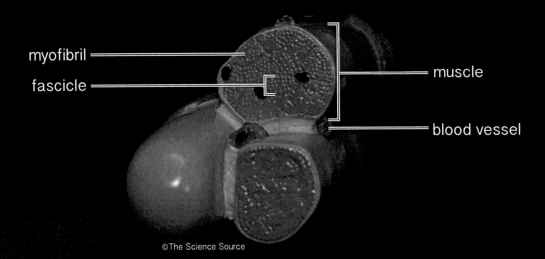

myofibril

fascicle

muscle

blood vessel

©The Science Source

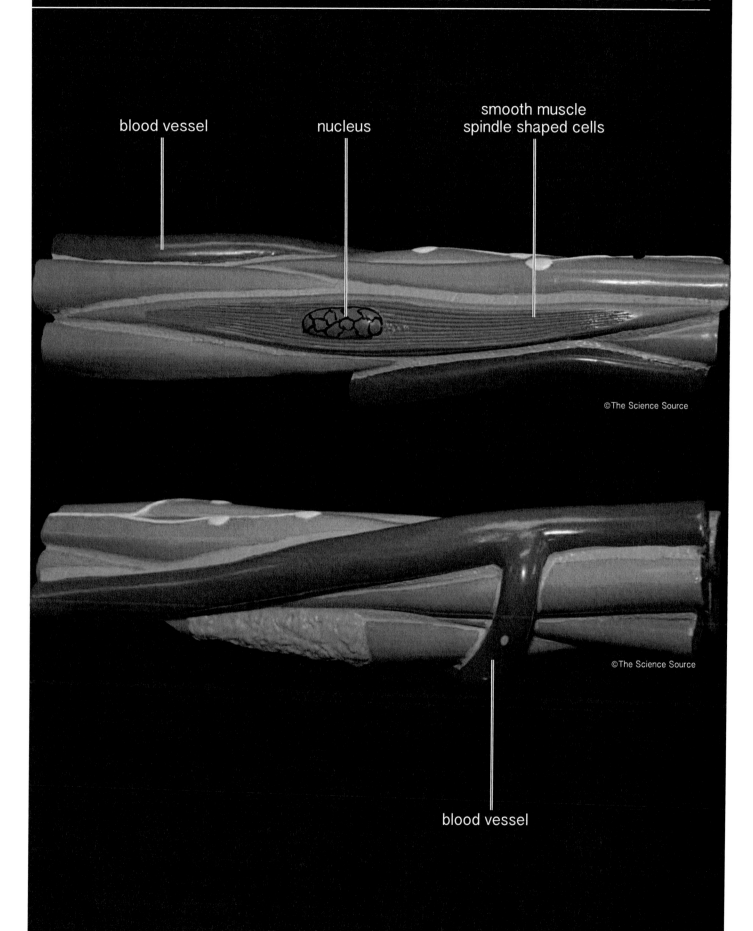

blood vessel

nucleus

smooth muscle
spindle shaped cells

©The Science Source

©The Science Source

blood vessel

Canine - CRANIAL MUSCLES

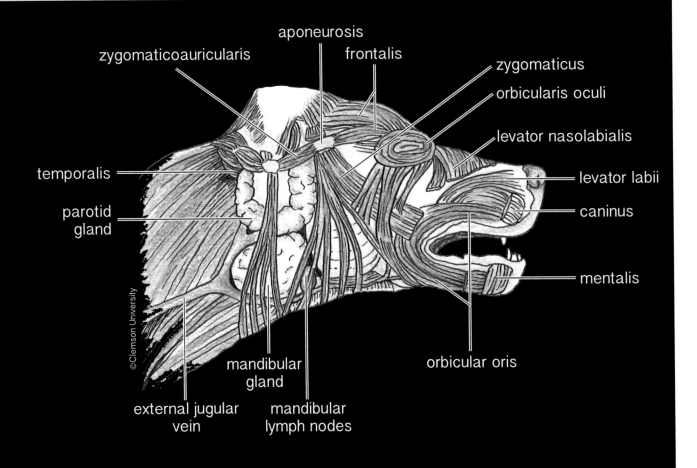

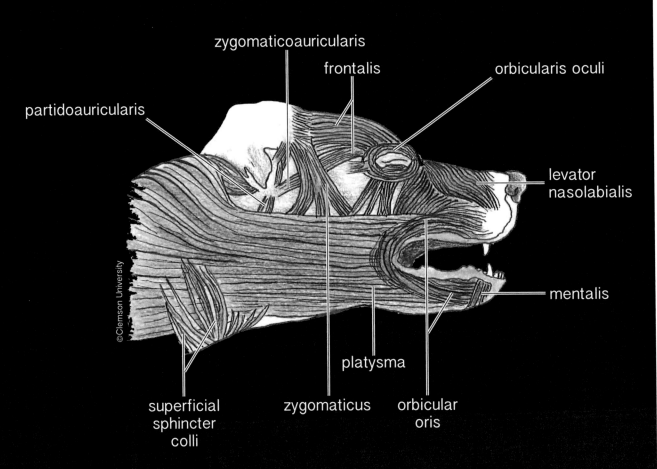

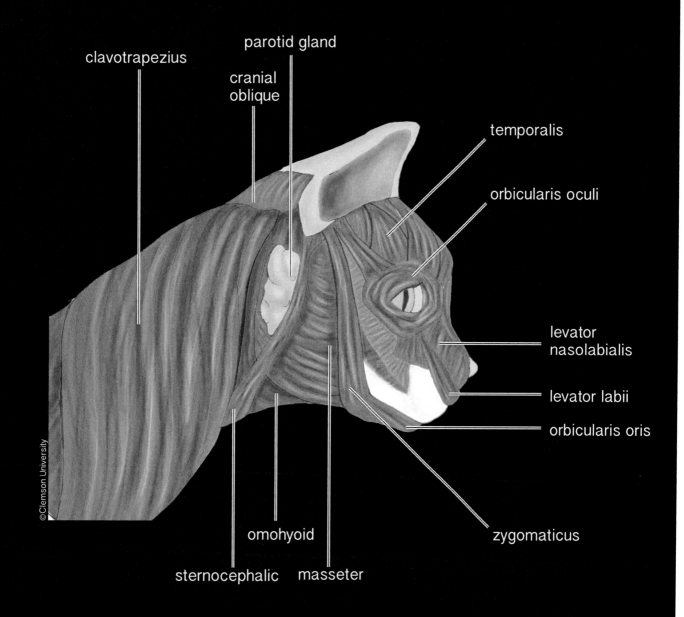

clavotrapezius

parotid gland

cranial
oblique

temporalis

orbicularis oculi

levator
nasolabialis

levator labii

orbicularis oris

zygomaticus

omohyoid

sternocephalic masseter

©Clemson University

FRONTAL MUSCLES

Canine

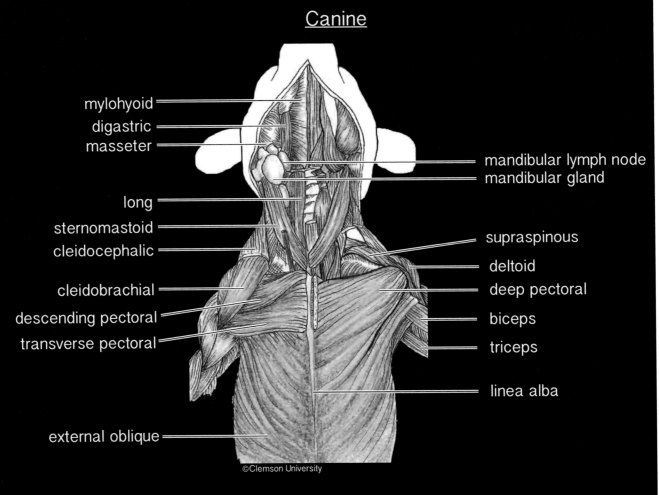

- mylohyoid
- digastric
- masseter
- mandibular lymph node
- mandibular gland
- long
- sternomastoid
- cleidocephalic
- supraspinous
- deltoid
- cleidobrachial
- deep pectoral
- descending pectoral
- biceps
- transverse pectoral
- triceps
- linea alba
- external oblique

©Clemson University

Feline

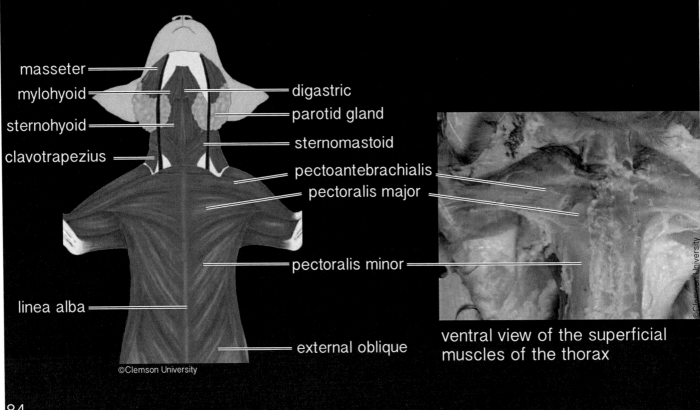

- masseter
- mylohyoid
- digastric
- parotid gland
- sternohyoid
- sternomastoid
- clavotrapezius
- pectoantebrachialis
- pectoralis major
- pectoralis minor
- linea alba
- external oblique

©Clemson University

ventral view of the superficial muscles of the thorax

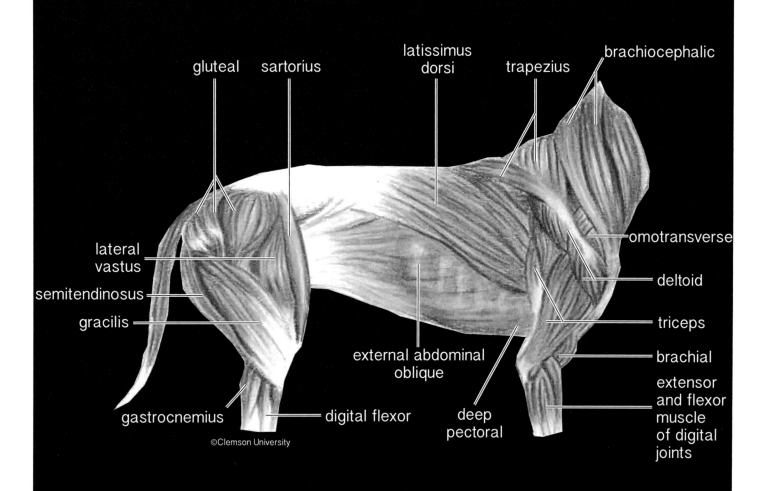

gluteal · sartorius · latissimus dorsi · trapezius · brachiocephalic

omotransverse

lateral vastus

deltoid

semitendinosus

gracilis

triceps

external abdominal oblique

brachial

gastrocnemius · digital flexor · deep pectoral

extensor and flexor muscle of digital joints

©Clemson University

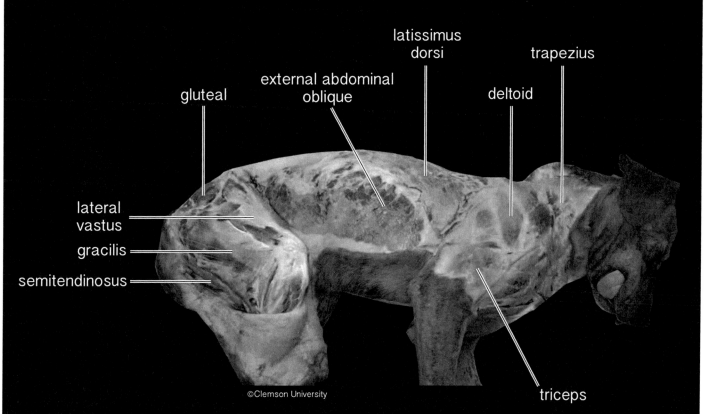

latissimus dorsi · trapezius

external abdominal oblique

deltoid

gluteal

lateral vastus

gracilis

semitendinosus

triceps

©Clemson University

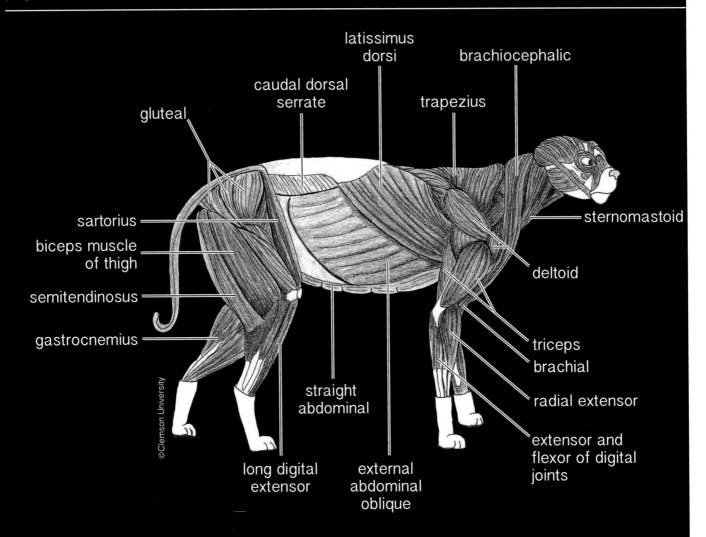

latissimus dorsi

caudal dorsal serrate

brachiocephalic

trapezius

gluteal

sartorius

sternomastoid

biceps muscle of thigh

deltoid

semitendinosus

gastrocnemius

triceps brachial

radial extensor

straight abdominal

long digital extensor

external abdominal oblique

extensor and flexor of digital joints

©Clemson University

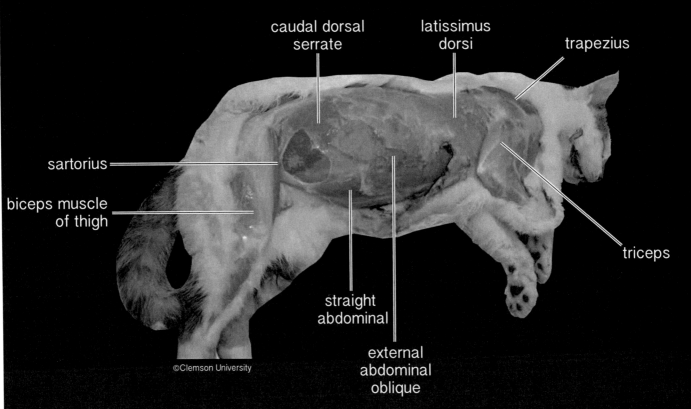

caudal dorsal serrate

latissimus dorsi

trapezius

sartorius

biceps muscle of thigh

triceps

straight abdominal

external abdominal oblique

©Clemson University

Nervous System/ Sensory Organs

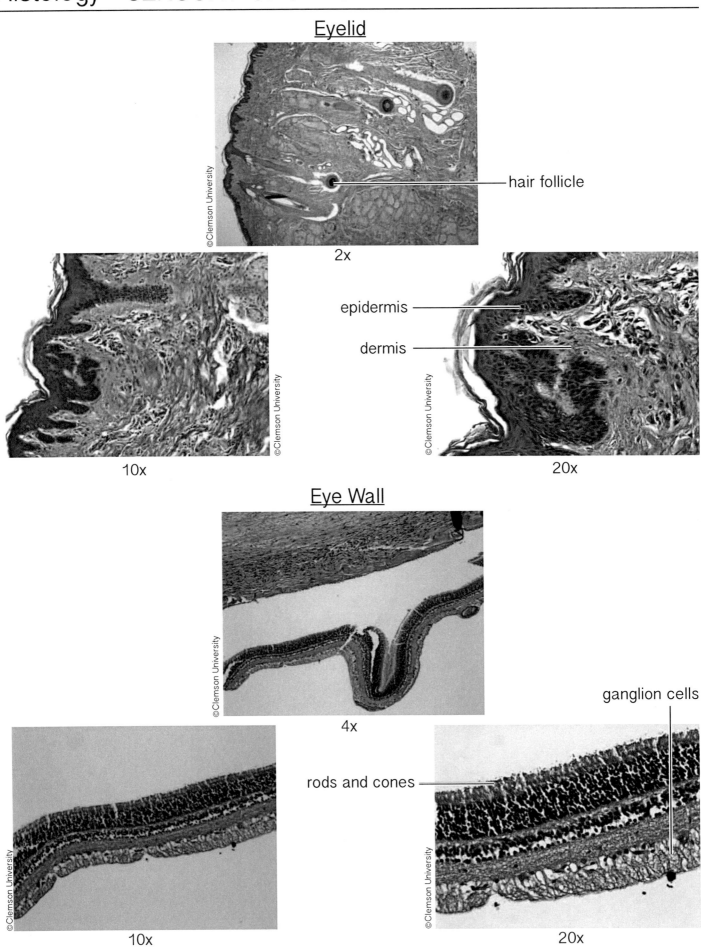

Eyelid

2x

— hair follicle

10x

epidermis

dermis

20x

Eye Wall

4x

10x

ganglion cells

rods and cones

20x

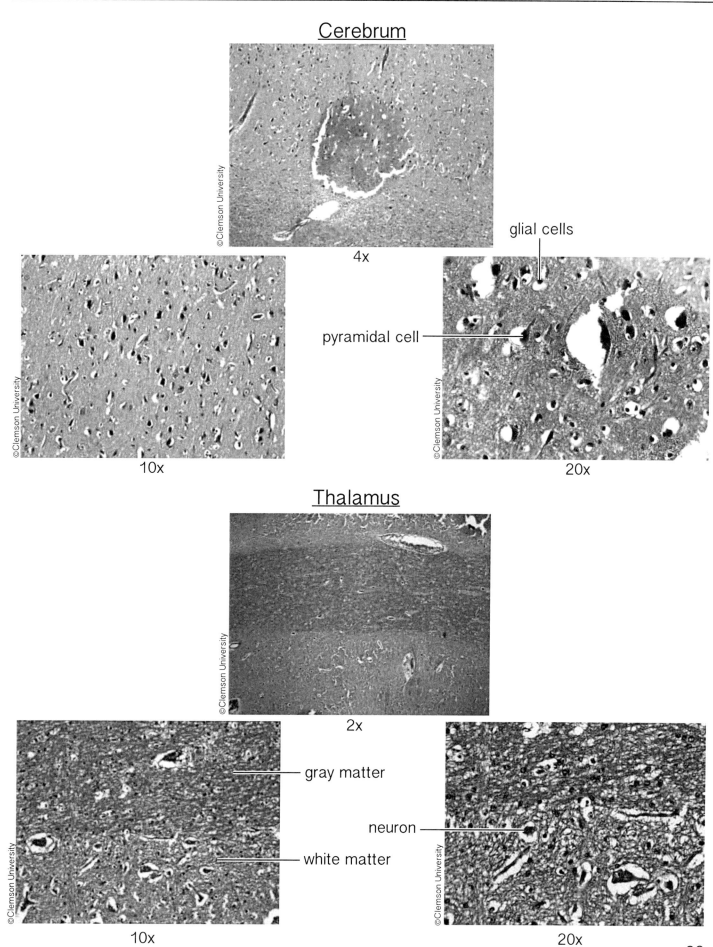

Cerebrum

4x

glial cells

pyramidal cell

10x

20x

Thalamus

2x

gray matter

white matter

neuron

10x

20x

89

Cerebellum

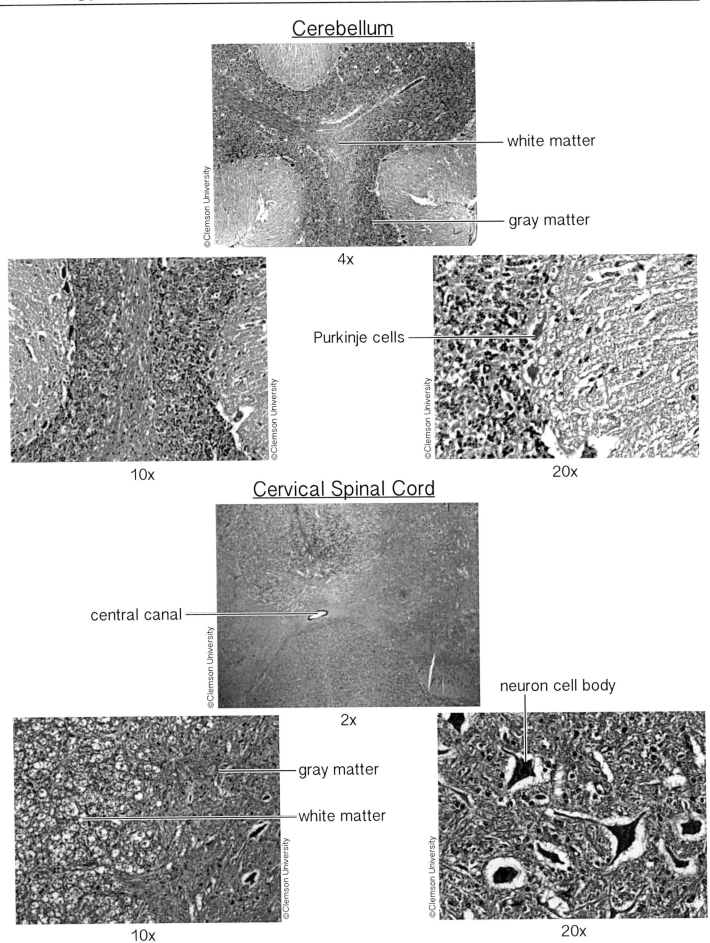

white matter

gray matter

4x

Purkinje cells

10x

20x

Cervical Spinal Cord

central canal

2x

neuron cell body

gray matter

white matter

10x

20x

<u>Thoracic Spinal Cord</u>

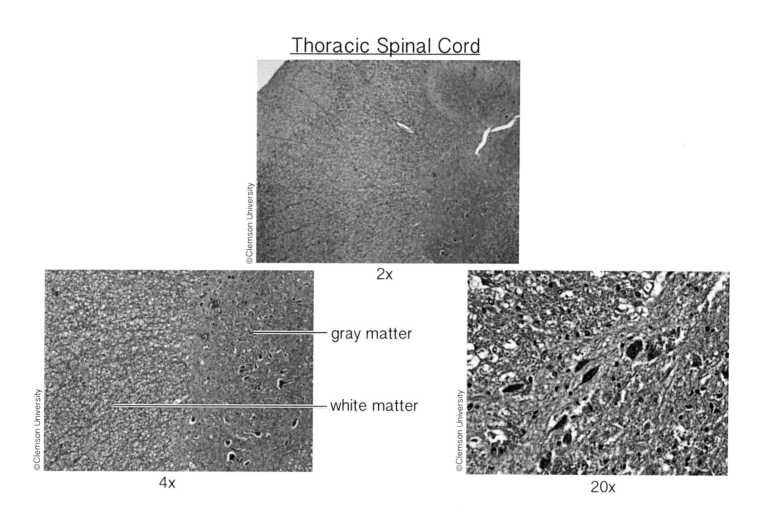

2x

gray matter

white matter

4x

20x

MAJOR NERVES

Canine

n. - nerve

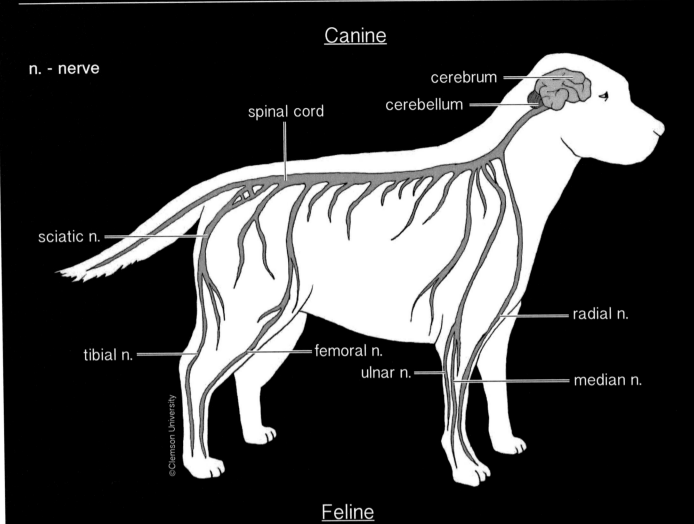

cerebrum
cerebellum
spinal cord
sciatic n.
radial n.
tibial n.
femoral n.
ulnar n.
median n.

©Clemson University

Feline

n. - nerve

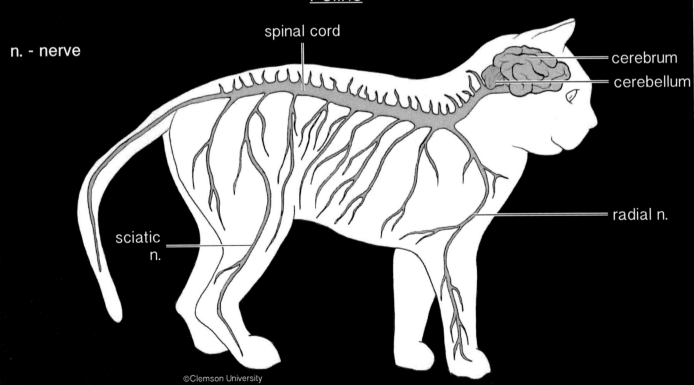

spinal cord
cerebrum
cerebellum
radial n.
sciatic n.

©Clemson University

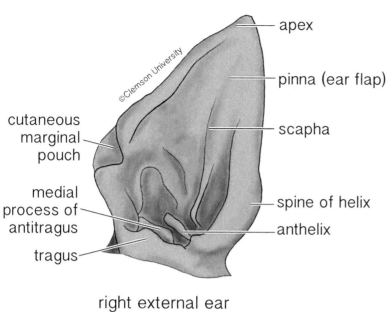

right external ear

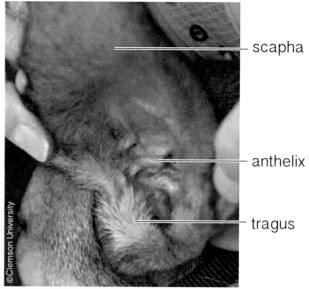

left external ear

Canine

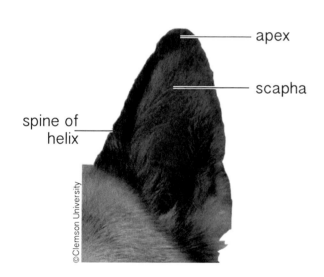

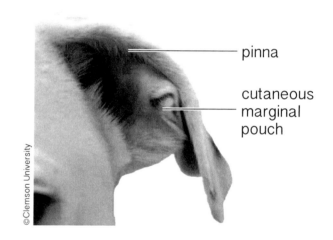

Feline

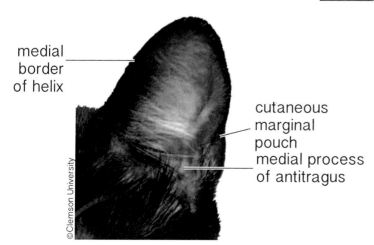

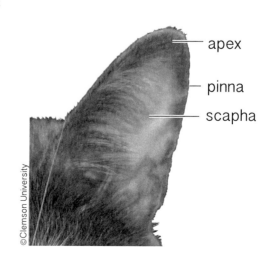

EARS

Healthy Ear

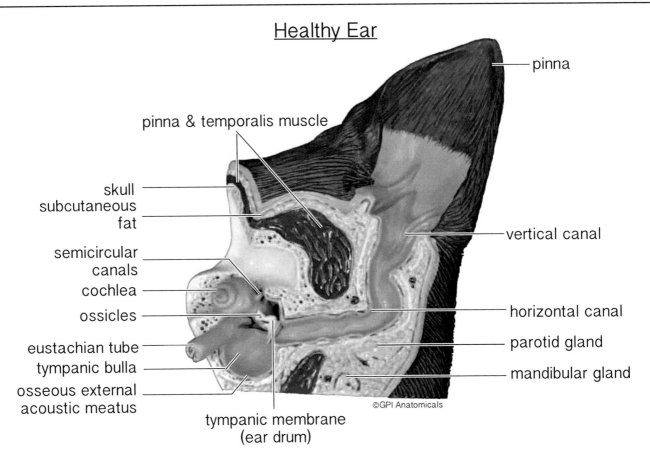

pinna

pinna & temporalis muscle

skull

subcutaneous fat

semicircular canals

cochlea

ossicles

eustachian tube

tympanic bulla

osseous external acoustic meatus

vertical canal

horizontal canal

parotid gland

mandibular gland

©GPI Anatomicals

tympanic membrane (ear drum)

Infected Ear

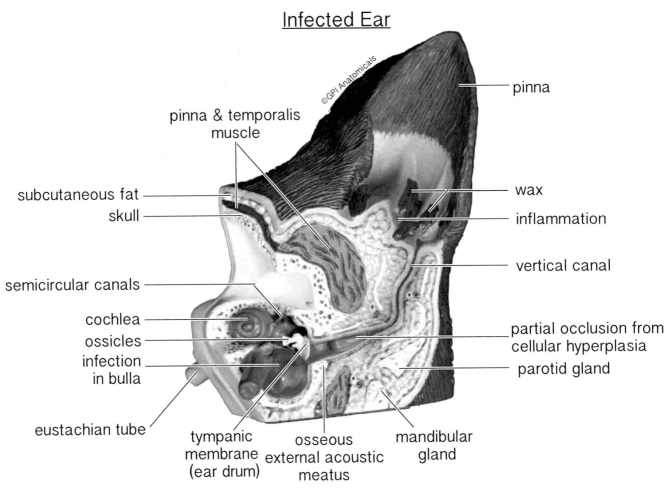

©GPI Anatomicals

pinna & temporalis muscle

subcutaneous fat

skull

semicircular canals

cochlea

ossicles

infection in bulla

eustachian tube

tympanic membrane (ear drum)

osseous external acoustic meatus

mandibular gland

pinna

wax

inflammation

vertical canal

partial occlusion from cellular hyperplasia

parotid gland

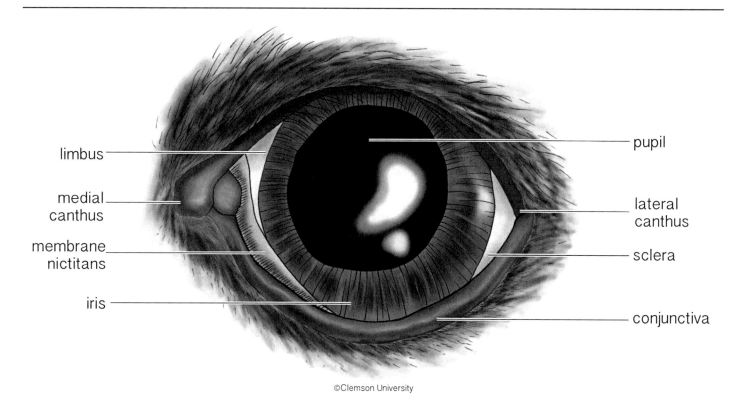

limbus

medial canthus

membrane nictitans

iris

pupil

lateral canthus

sclera

conjunctiva

©Clemson University

Canine

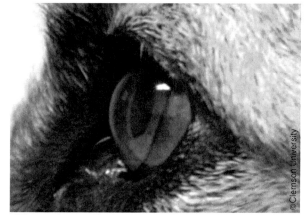

dilated pupil, lateral view

dilated pupil

Feline

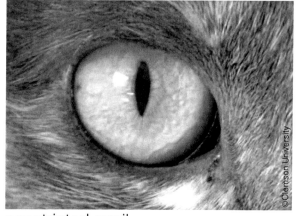

constricted pupil

dilated pupil

Circulatory/ Lymphatic System

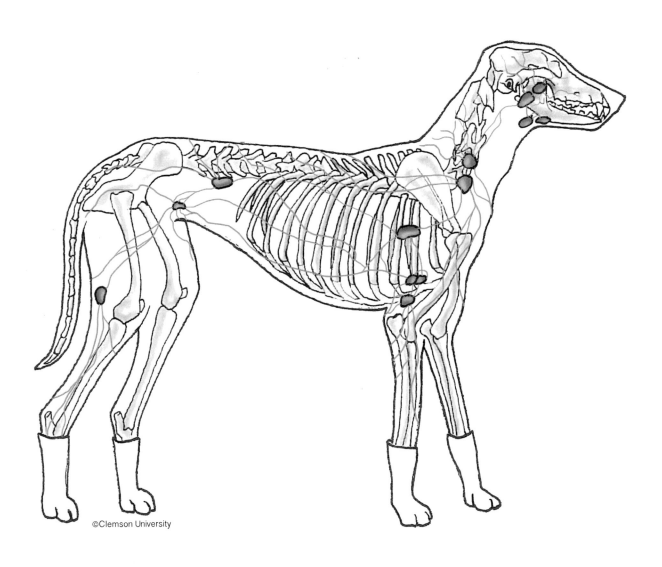

Cardiac Muscle

4x

10x

nuclei

muscle fibers

20x

Atrioventricular Valve

atrioventricular valve

atrial wall

4x

10x

20x

97

Aorta

tunica intima

tunica media

tunica adventitia

2x

4x

tunica media
- collagenous fibers
- elastic lamellae
- smooth muscle cells

20x

Vena Cava

2x

4x

tunica media

tunica intima

20x

Blood

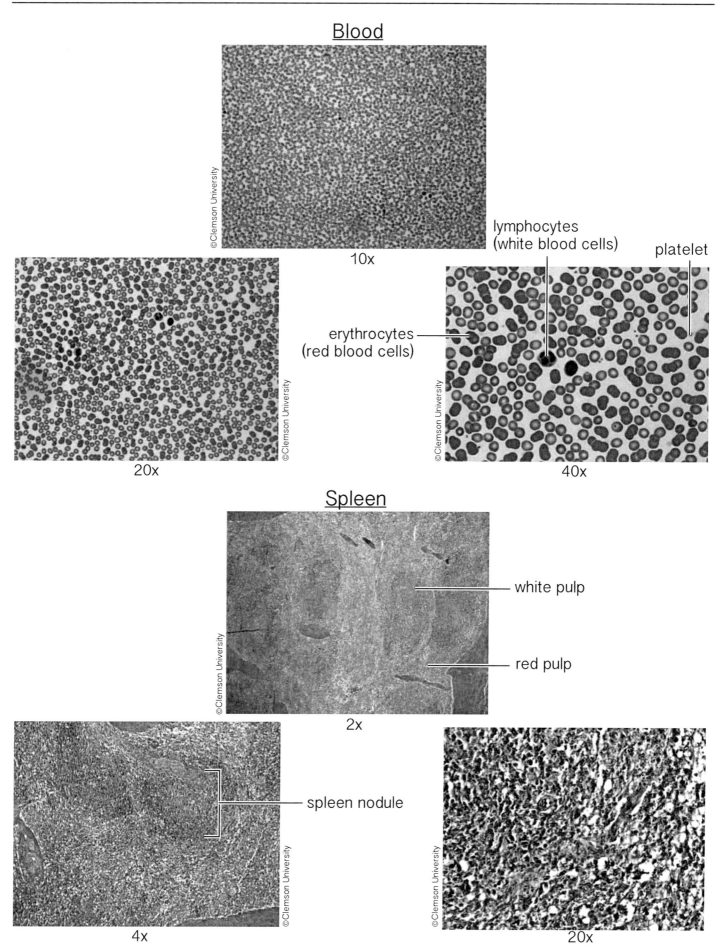

10x

20x

erythrocytes
(red blood cells)

lymphocytes
(white blood cells)

platelet

40x

Spleen

white pulp

red pulp

2x

spleen nodule

4x

20x

Histology - LYMPHATIC SYSTEM

Tonsil

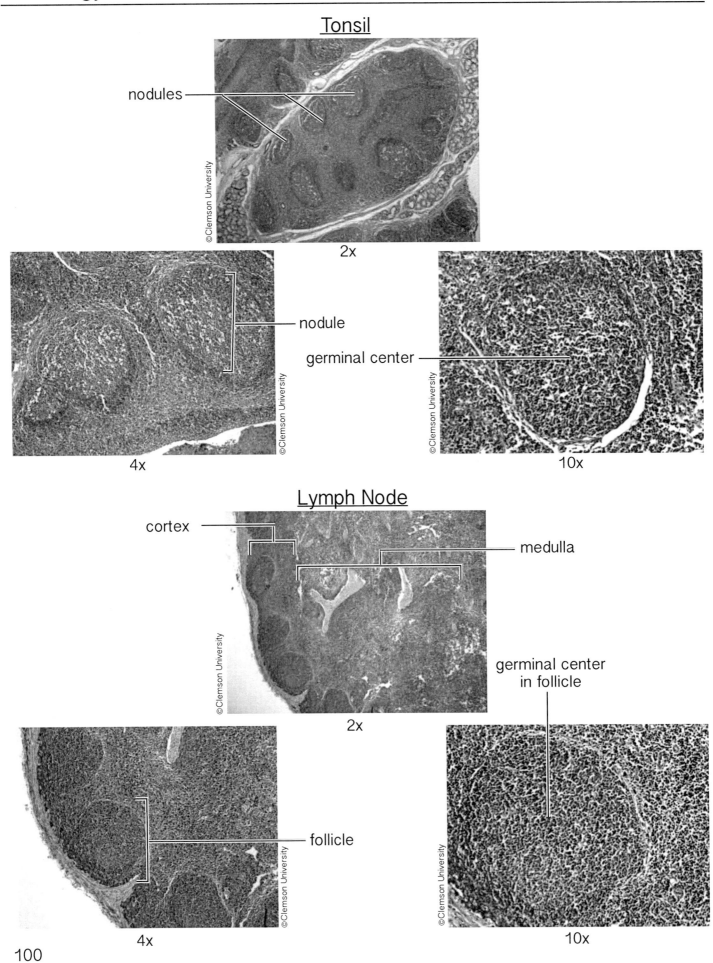

nodules

2x

nodule

germinal center

4x

10x

Lymph Node

cortex

medulla

2x

germinal center
in follicle

follicle

4x

10x

Arteries and Veins

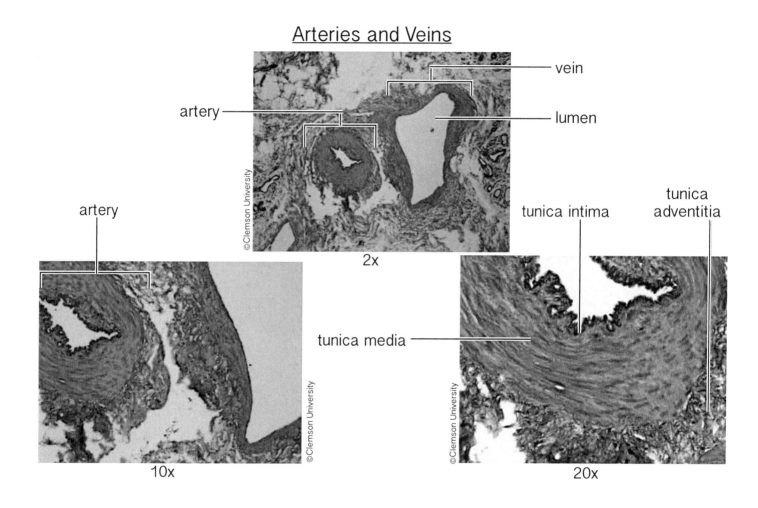

VEINS

Canine

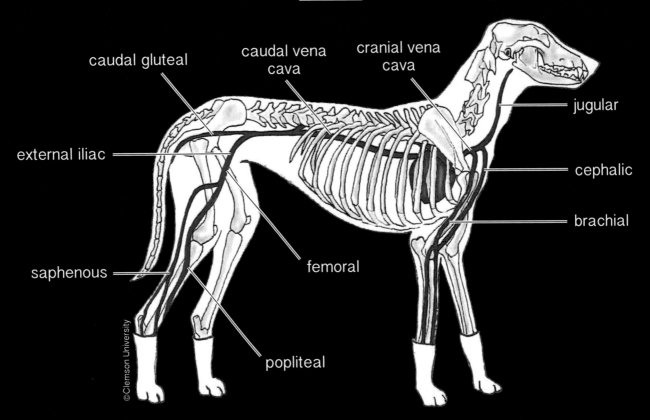

caudal gluteal

caudal vena cava

cranial vena cava

jugular

external iliac

cephalic

brachial

saphenous

femoral

popliteal

©Clemson University

Feline

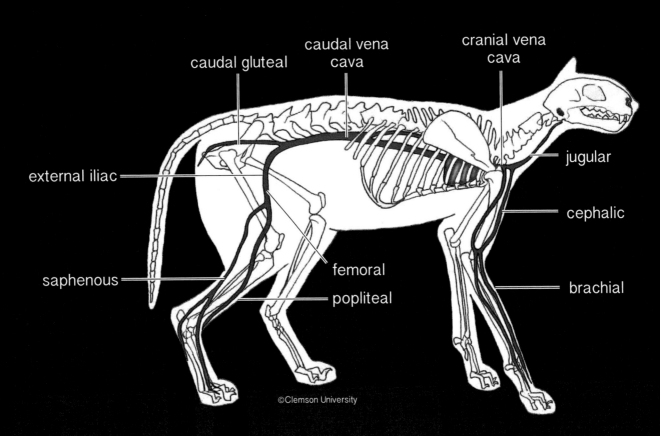

caudal gluteal

caudal vena cava

cranial vena cava

jugular

external iliac

cephalic

saphenous

femoral

brachial

popliteal

©Clemson University

Canine

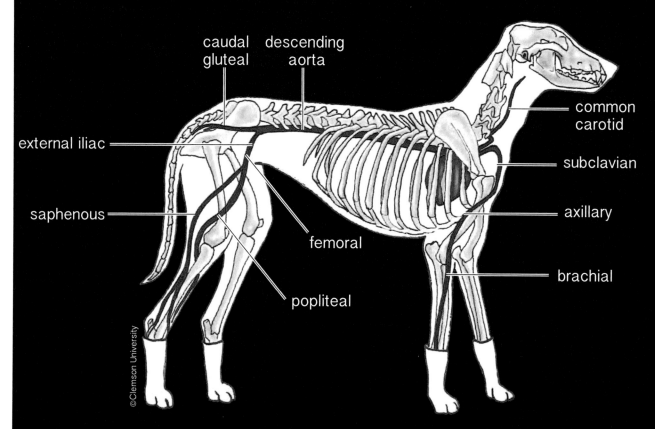

caudal gluteal
descending aorta
common carotid
external iliac
subclavian
saphenous
axillary
femoral
brachial
popliteal

©Clemson University

Feline

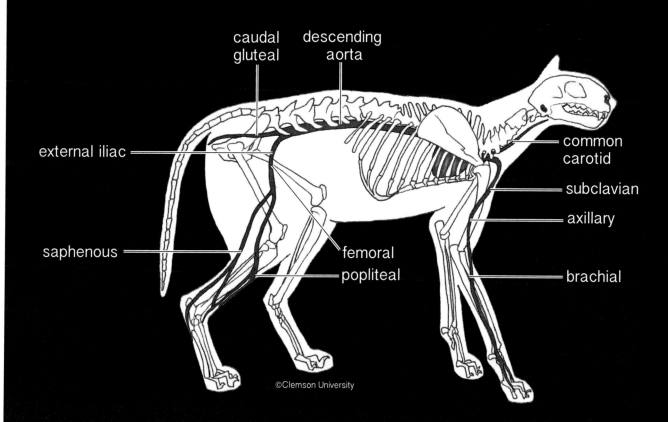

caudal gluteal
descending aorta
common carotid
external iliac
subclavian
axillary
femoral
saphenous
popliteal
brachial

©Clemson University

MAJOR CIRCULATORY VESSELS

Feline

Artery - a.
Vein - v.

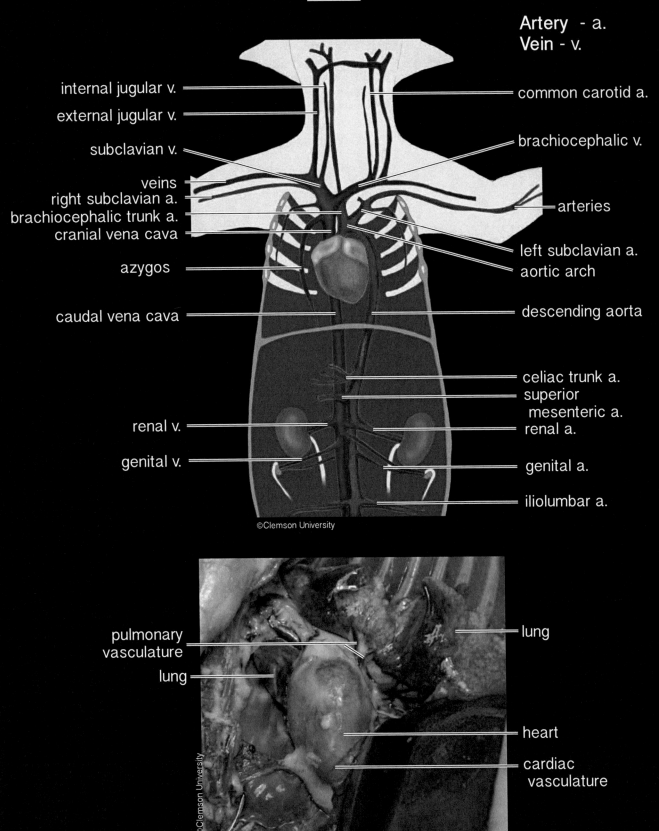

internal jugular v.

external jugular v.

subclavian v.

veins

right subclavian a.

brachiocephalic trunk a.

cranial vena cava

azygos

caudal vena cava

renal v.

genital v.

common carotid a.

brachiocephalic v.

arteries

left subclavian a.

aortic arch

descending aorta

celiac trunk a.

superior mesenteric a.

renal a.

genital a.

iliolumbar a.

©Clemson University

pulmonary vasculature

lung

lung

heart

cardiac vasculature

©Clemson University

Heartworm

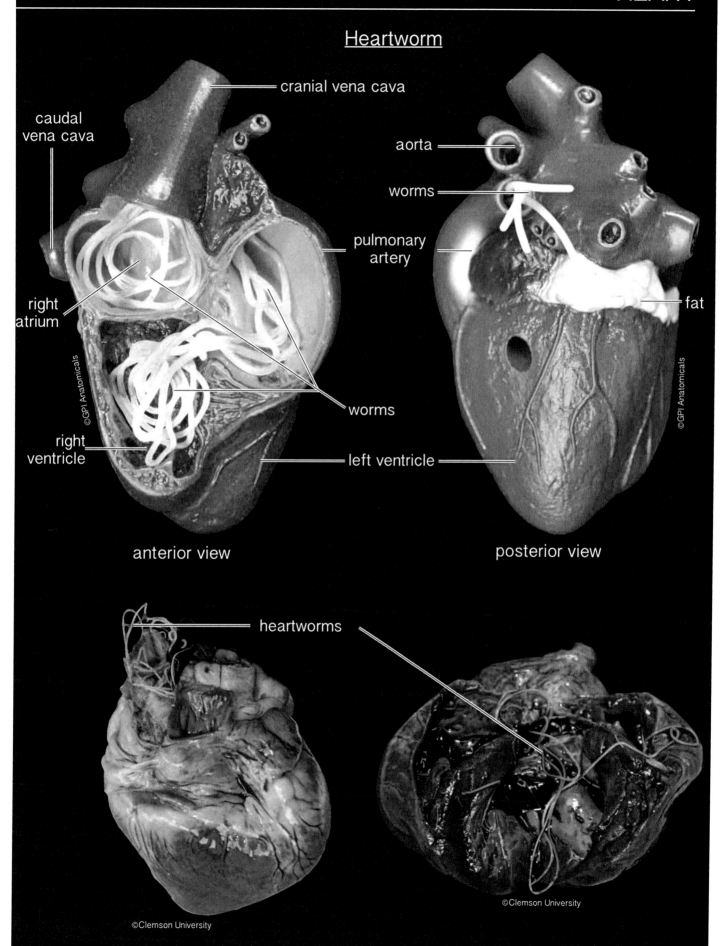

cranial vena cava

caudal vena cava

aorta

worms

pulmonary artery

right atrium

fat

©GPI Anatomicals

worms

right ventricle

left ventricle

©GPI Anatomicals

anterior view

posterior view

heartworms

©Clemson University

©Clemson University

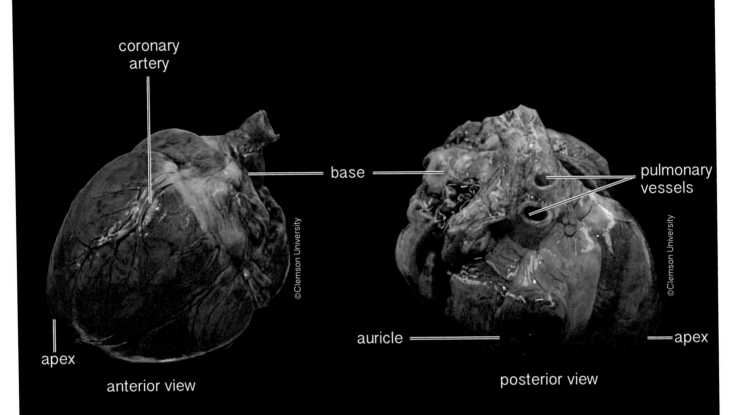

coronary artery

base

pulmonary vessels

©Clemson University

©Clemson University

apex

anterior view

auricle

apex

posterior view

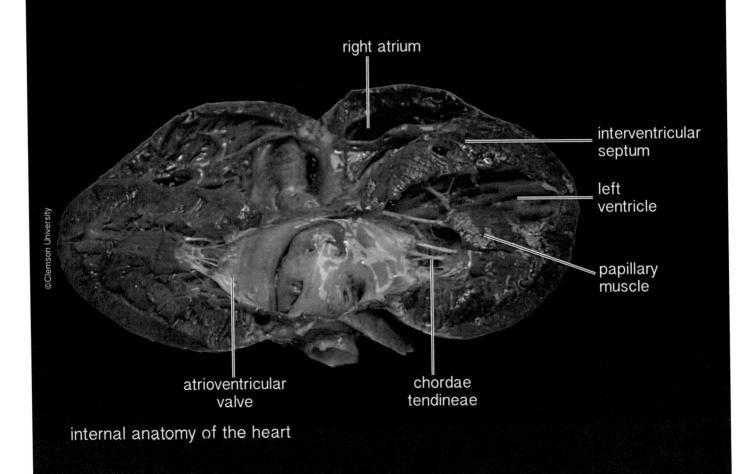

right atrium

interventricular septum

left ventricle

©Clemson University

papillary muscle

atrioventricular valve

chordae tendineae

internal anatomy of the heart

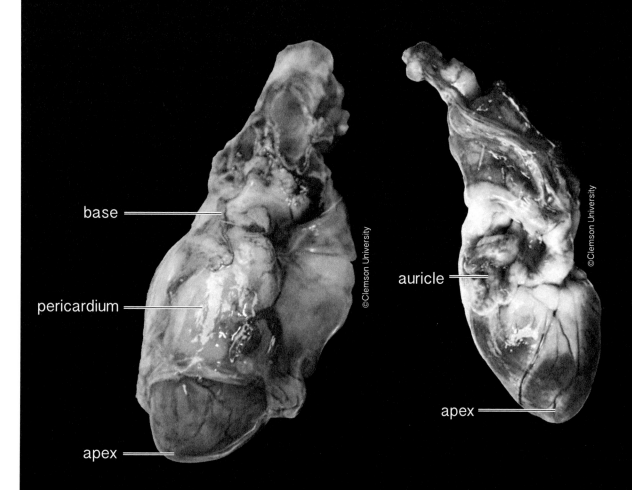

base

pericardium

apex

auricle

apex

©Clemson University

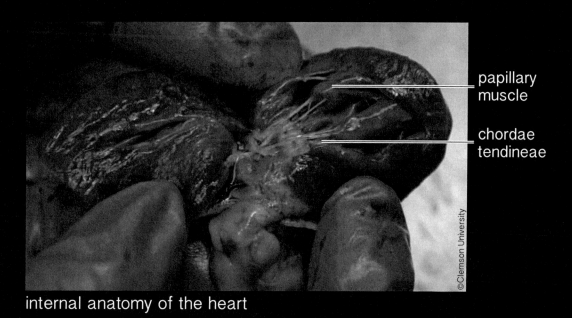

papillary muscle

chordae tendineae

©Clemson University

internal anatomy of the heart

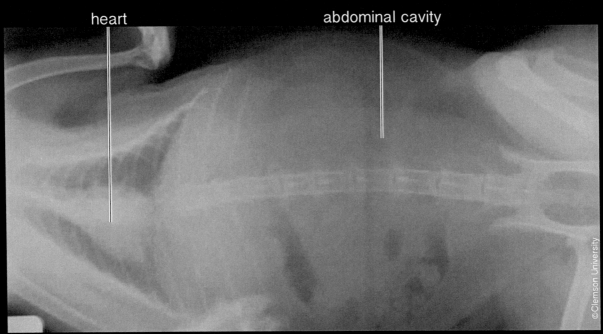

heart   abdominal cavity

ventrodorsal aspect

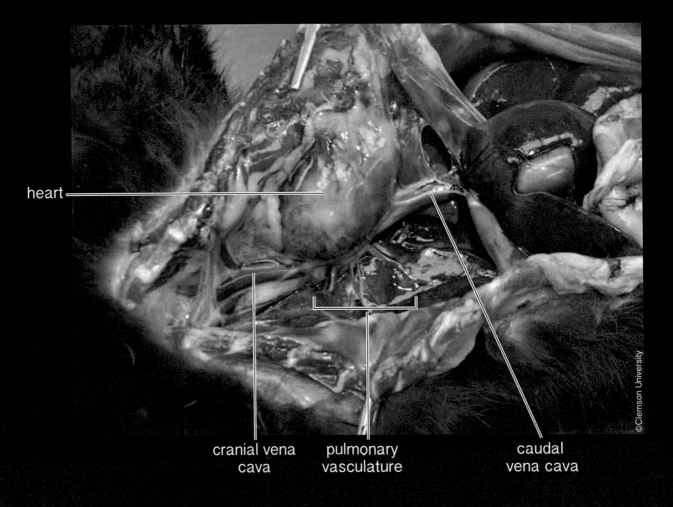

heart

cranial vena cava   pulmonary vasculature   caudal vena cava

Canine

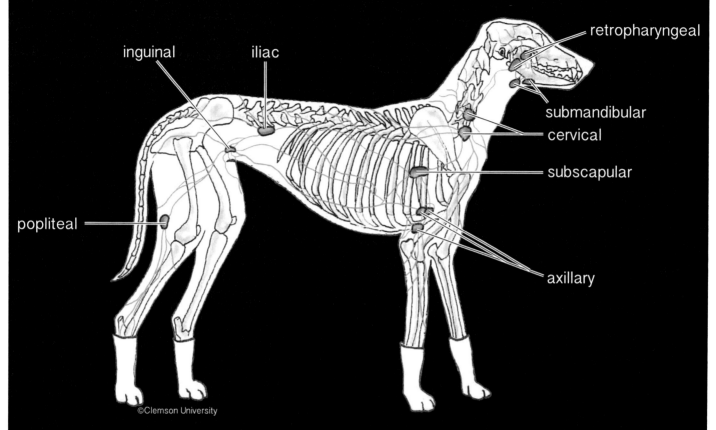

Feline

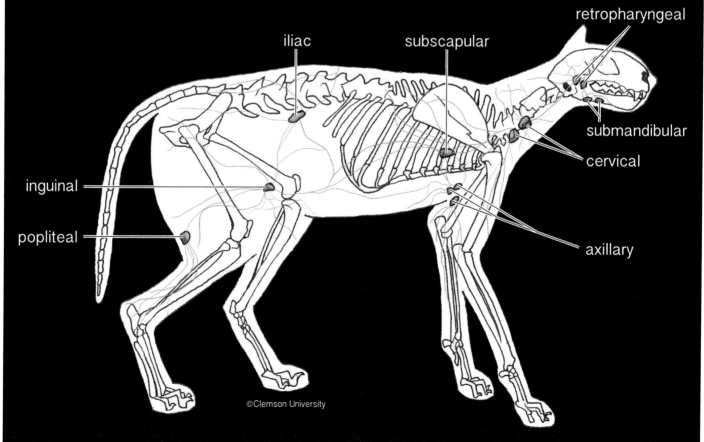

SPLEEN

Canine

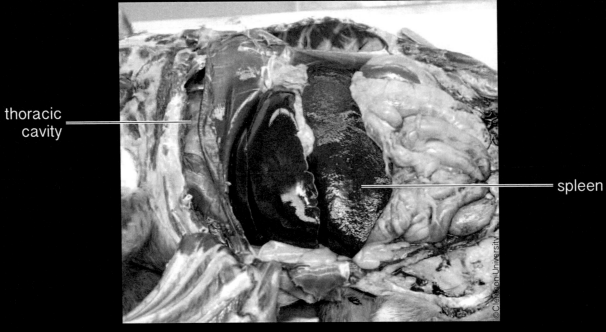

thoracic cavity

spleen

spleen in abdominal cavity

Feline

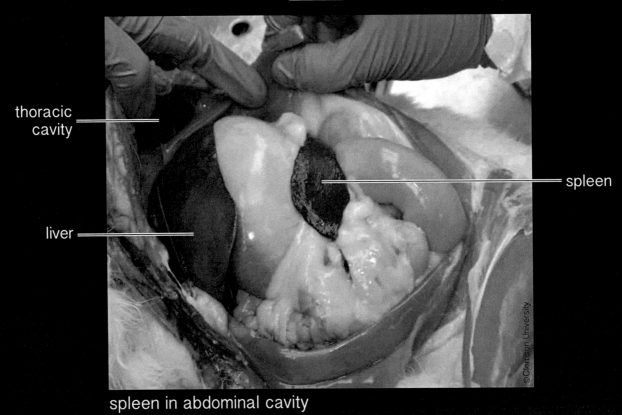

thoracic cavity

liver

spleen

spleen in abdominal cavity

Respiratory System

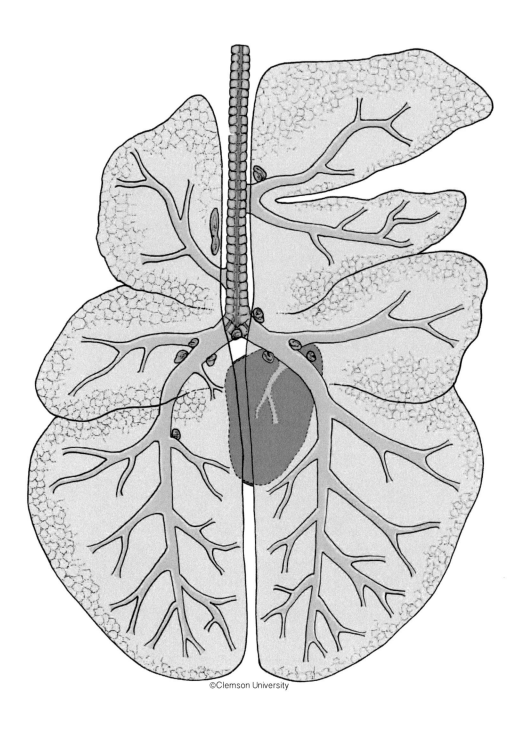

Histology - RESPIRATORY SYSTEM

Trachea

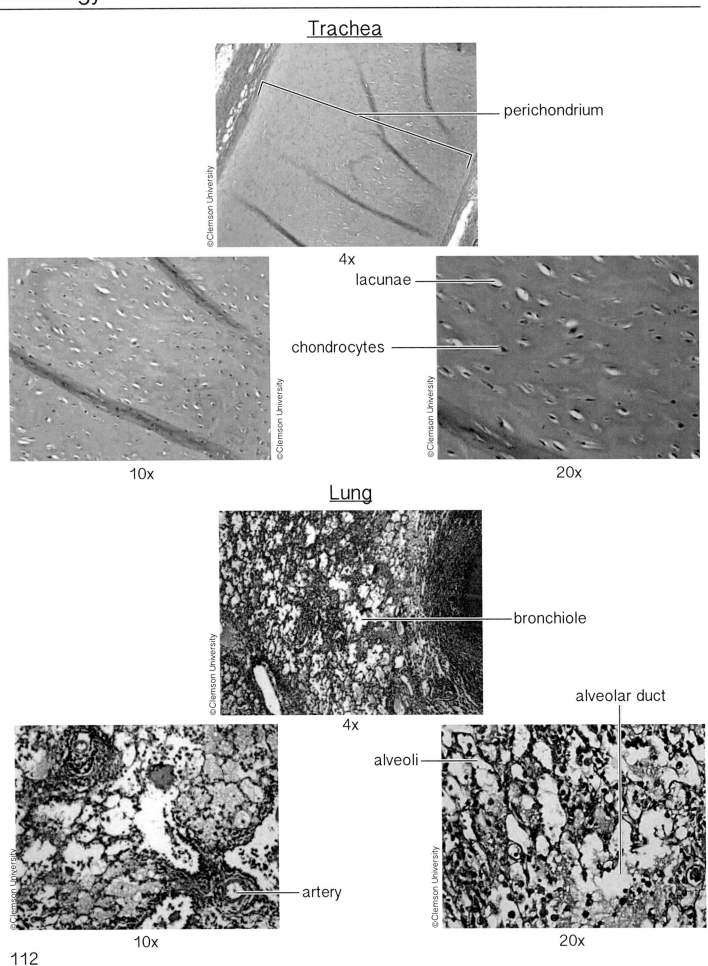

perichondrium

4x

lacunae

chondrocytes

10x

20x

Lung

bronchiole

alveolar duct

alveoli

artery

4x

10x

20x

Canine

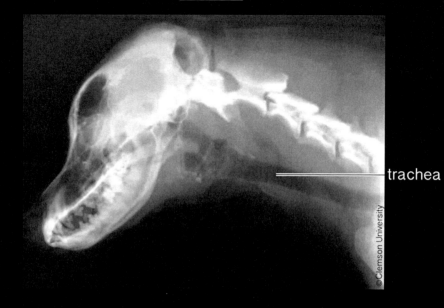

trachea

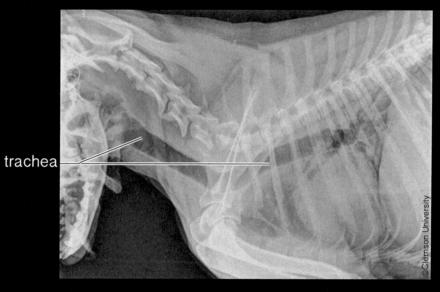

trachea

Feline

trachea

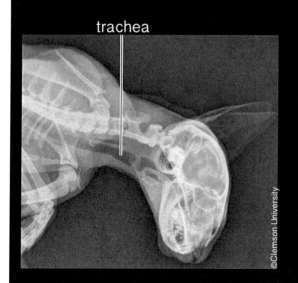

esophagus   trachea

LUNGS

Dorsal aspect of lungs

trachea

lymph node

left cranial lobe
(cranial)

left cranial lobe
(caudal)

left caudal
lobe

right cranial
lobe (cranial)

right cranial
lobe (caudal)

middle lobe

bronchus

accessory
lobe

right caudal
lobe

©Clemson University

Bronchial tree

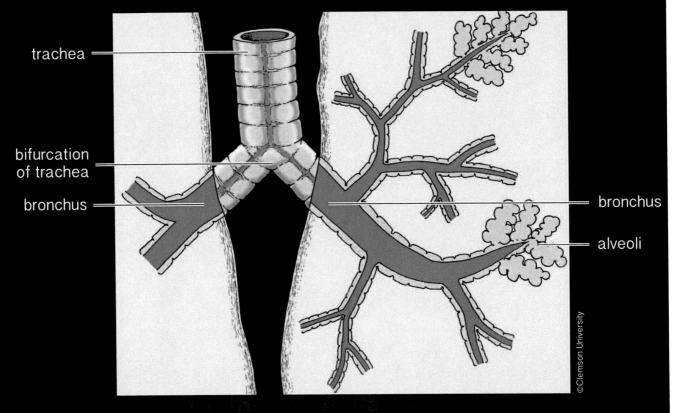

trachea

bifurcation
of trachea

bronchus

bronchus

alveoli

©Clemson University

Feline

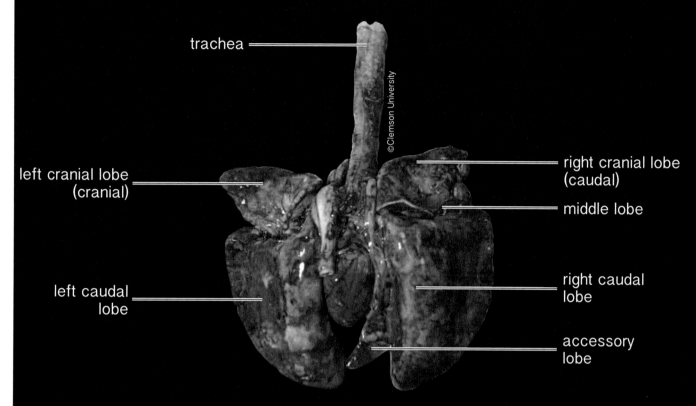

trachea

left cranial lobe
(cranial)

left caudal
lobe

right cranial lobe
(caudal)

middle lobe

right caudal
lobe

accessory
lobe

©Clemson University

trachea and lungs, dorsal aspect

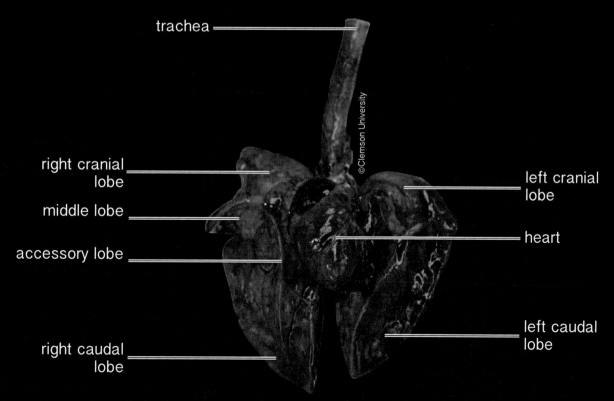

trachea

right cranial
lobe

middle lobe

accessory lobe

right caudal
lobe

left cranial
lobe

heart

left caudal
lobe

©Clemson University

trachea and lungs, ventral aspect

DIAPHRAGM

Canine

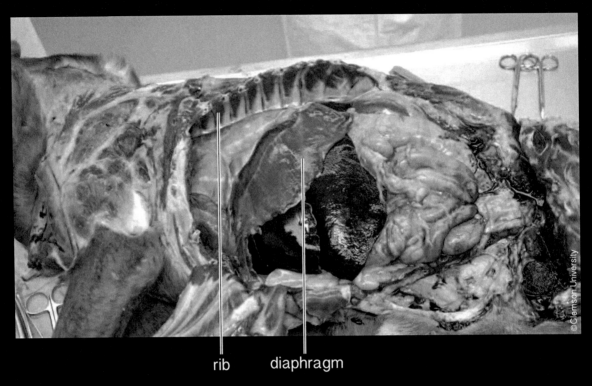

rib   diaphragm

Feline

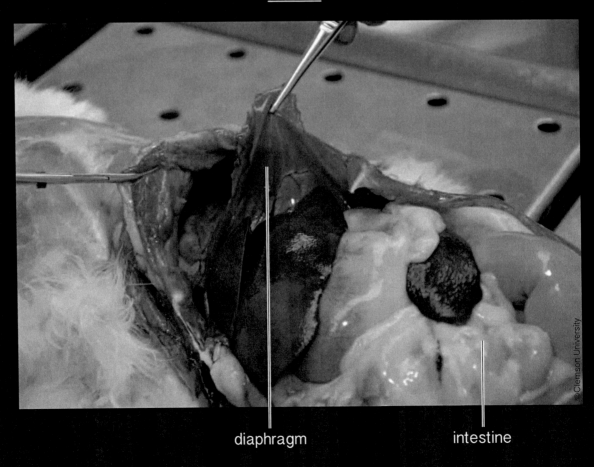

diaphragm   intestine

Chapter 8

Digestive System

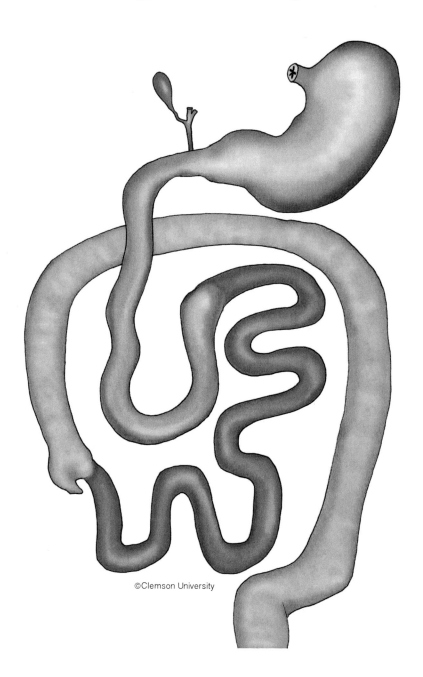

©Clemson University

Gingiva

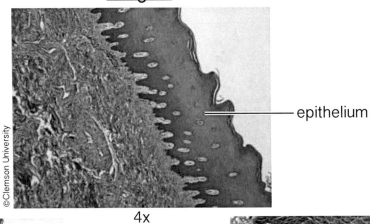

epithelium

4x

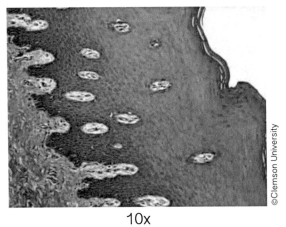

10x

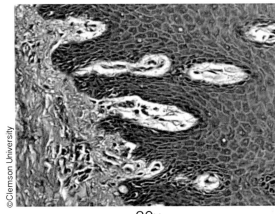

20x

Tongue

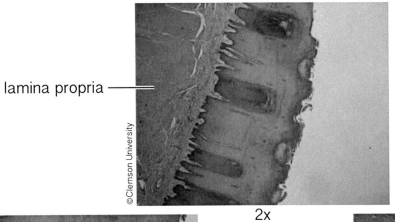

lamina propria

2x

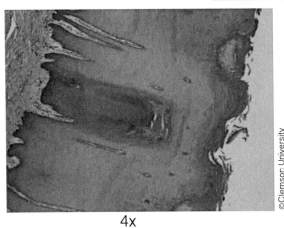

4x

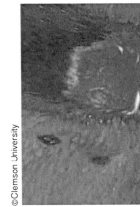

nonkeratinized stratified squamous epithelium

10x

Fungiform Papillae

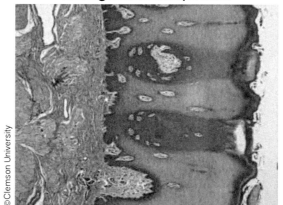

4x

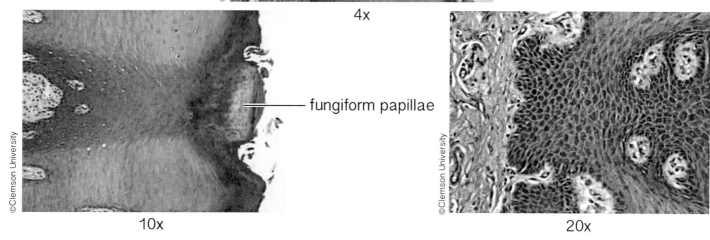

—— fungiform papillae

10x   20x

Circumvallate Papillae

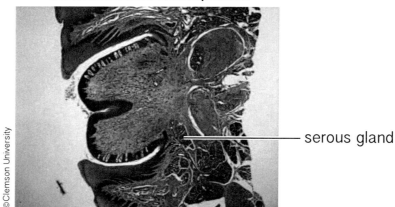

—— serous gland

2x

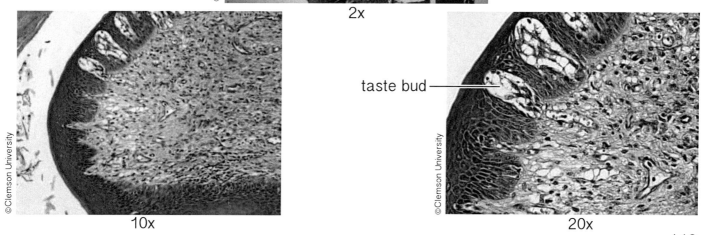

taste bud ——

10x   20x

Histology - DIGESTIVE SYSTEM

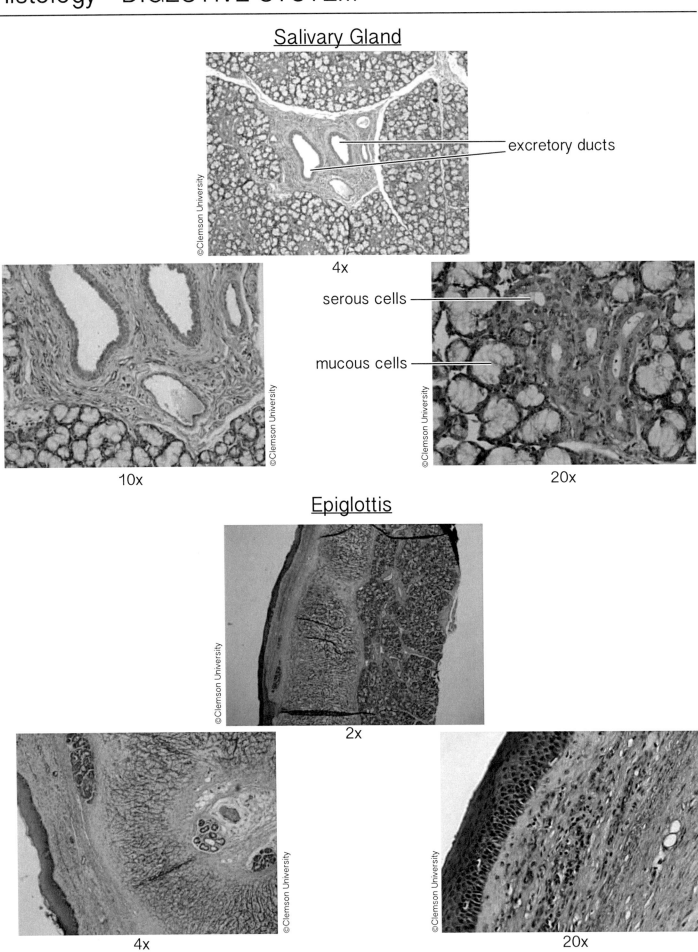

Salivary Gland

excretory ducts

4x

serous cells

mucous cells

10x

20x

Epiglottis

2x

4x

20x

Esophagus

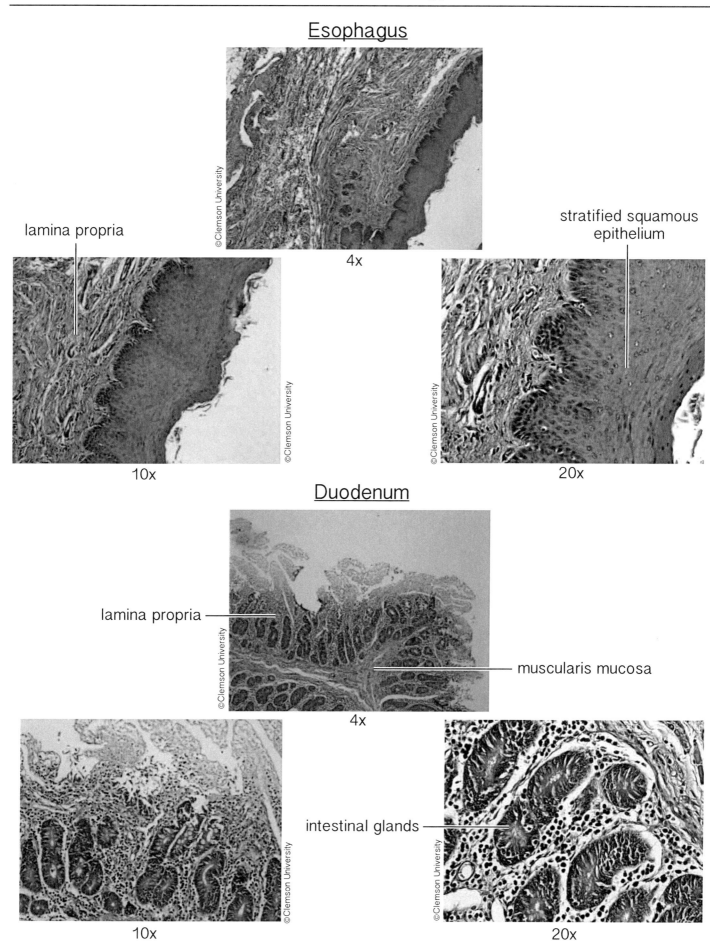

lamina propria

stratified squamous epithelium

4x

10x

20x

Duodenum

lamina propria

muscularis mucosa

4x

intestinal glands

10x

20x

121

Histology - DIGESTIVE SYSTEM

Jejunum

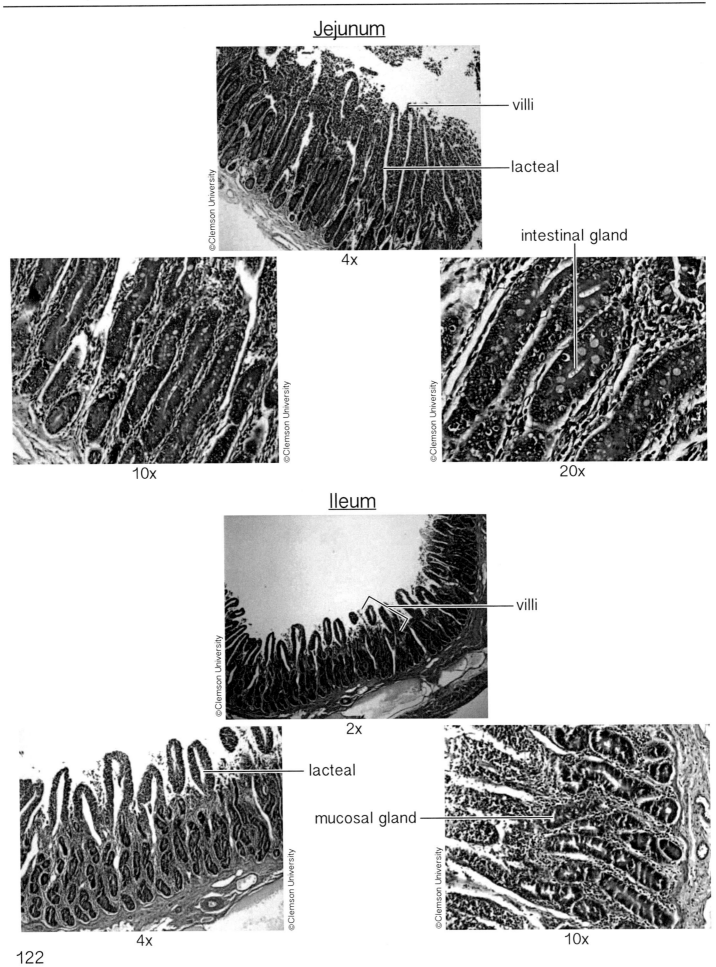

villi

lacteal

intestinal gland

©Clemson University

4x

10x

20x

Ileum

©Clemson University

2x

lacteal

mucosal gland

4x

10x

Ileocecal Junction

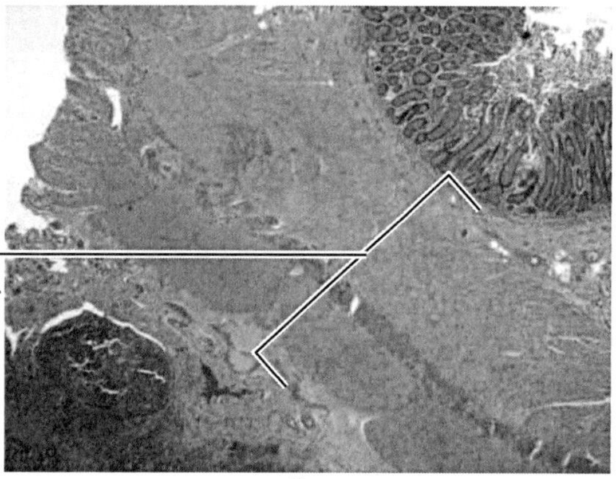

ileocecal junction

2x

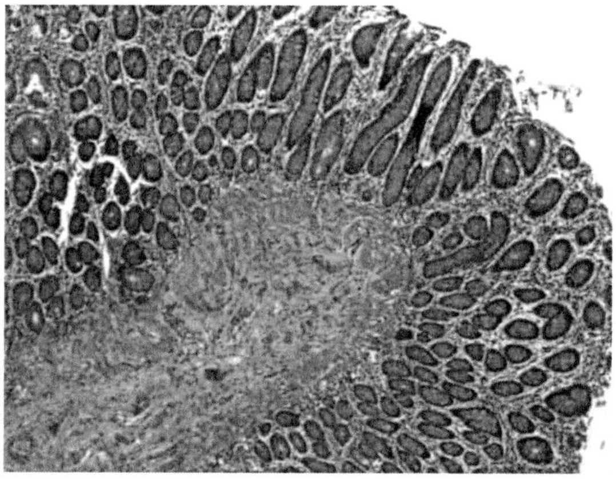

4x

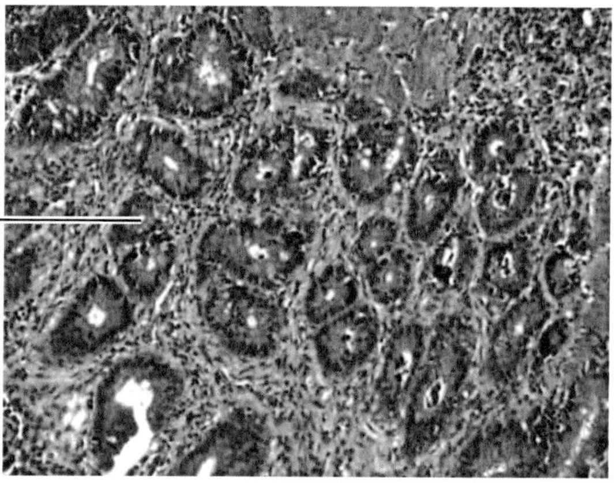

crypts

10x

Cecum

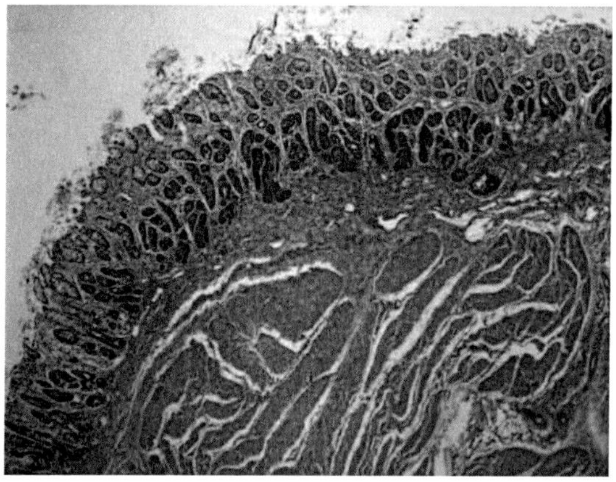

2x

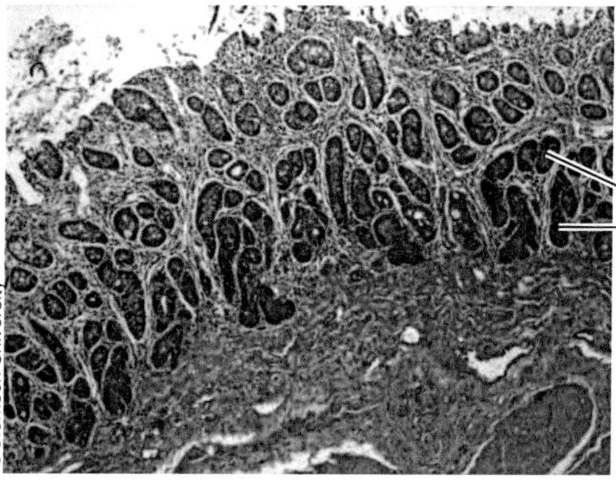

crypts

4x

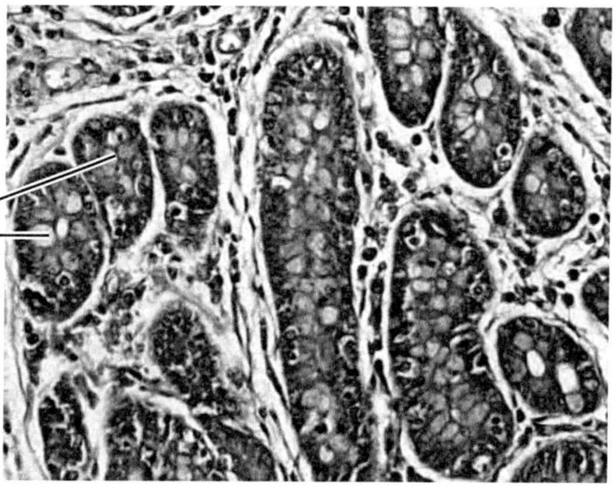

crypts

20x

123

Colon

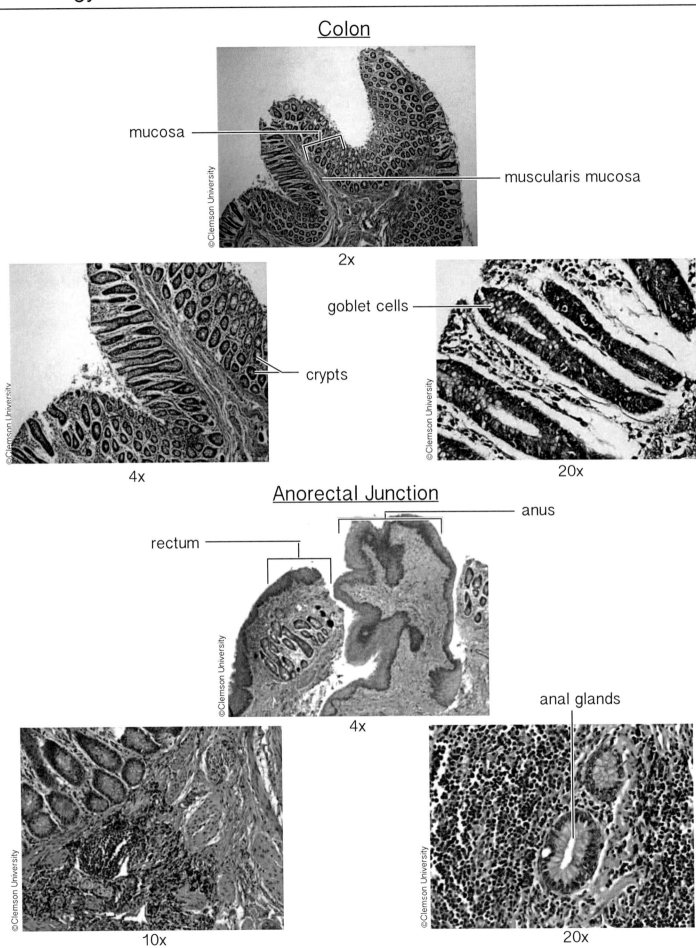

mucosa

muscularis mucosa

2x

goblet cells

crypts

4x

20x

Anorectal Junction

anus

rectum

©Clemson University

4x

anal glands

10x

20x

Liver

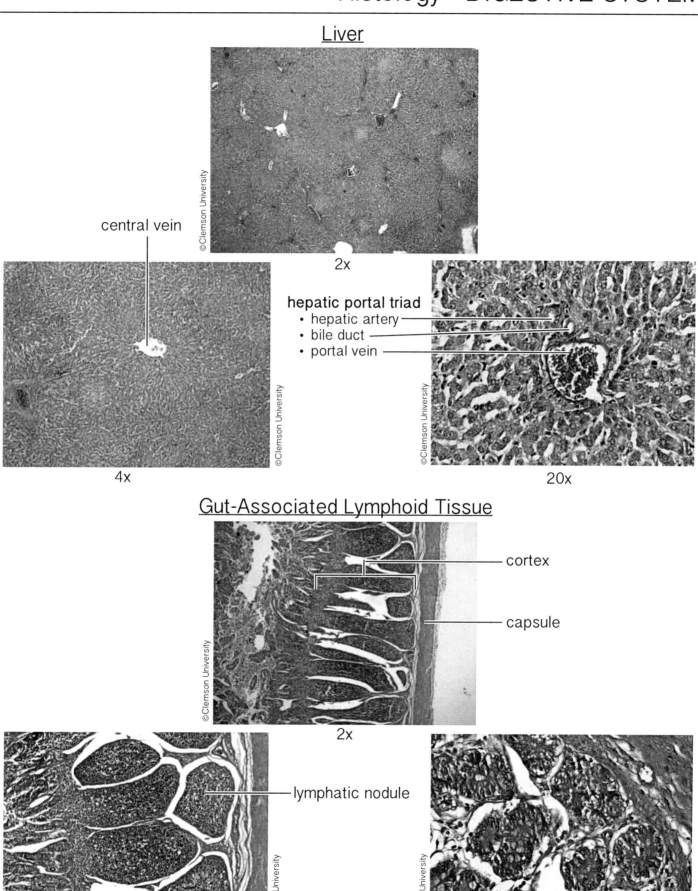

central vein

2x

hepatic portal triad
- hepatic artery
- bile duct
- portal vein

4x

20x

Gut-Associated Lymphoid Tissue

cortex

capsule

2x

lymphatic nodule

4x

20x

Omental Fat

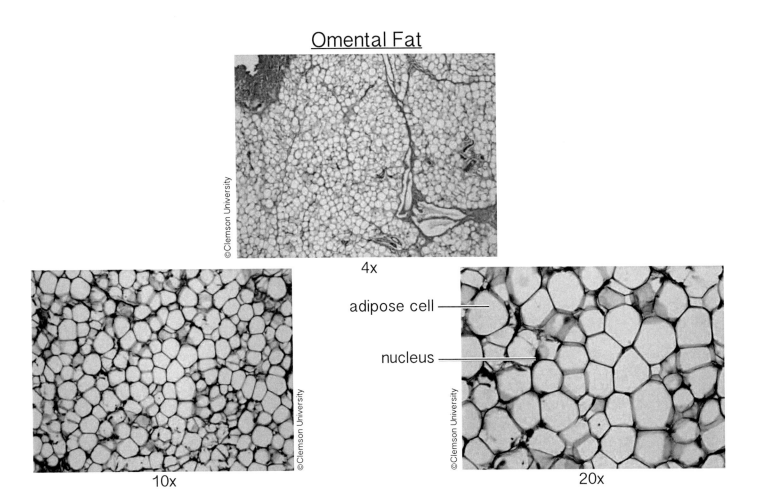

4x

adipose cell

nucleus

10x

20x

Canine

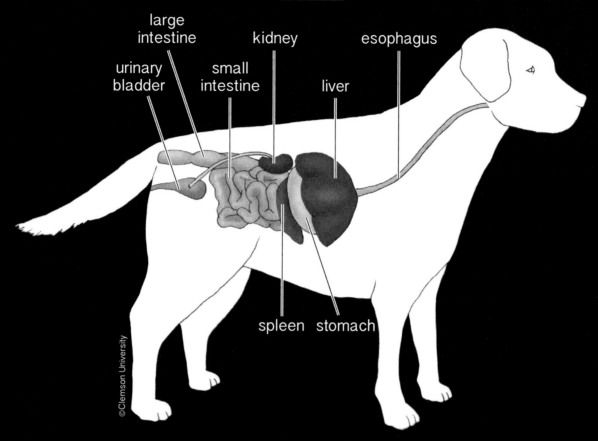

large intestine · kidney · esophagus · urinary bladder · small intestine · liver · spleen · stomach · ©Clemson University

Feline

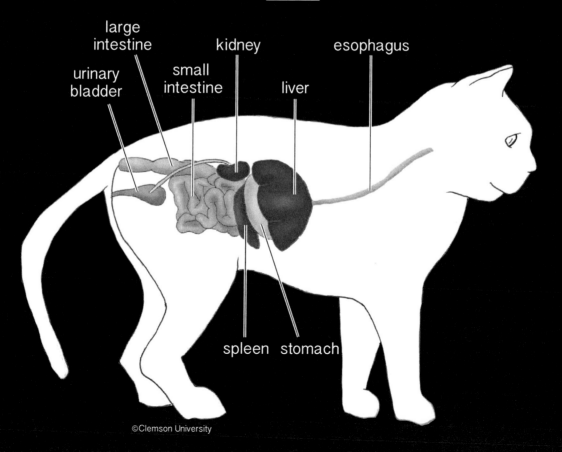

large intestine · kidney · esophagus · urinary bladder · small intestine · liver · spleen · stomach

GASTROINTESTINAL TRACT

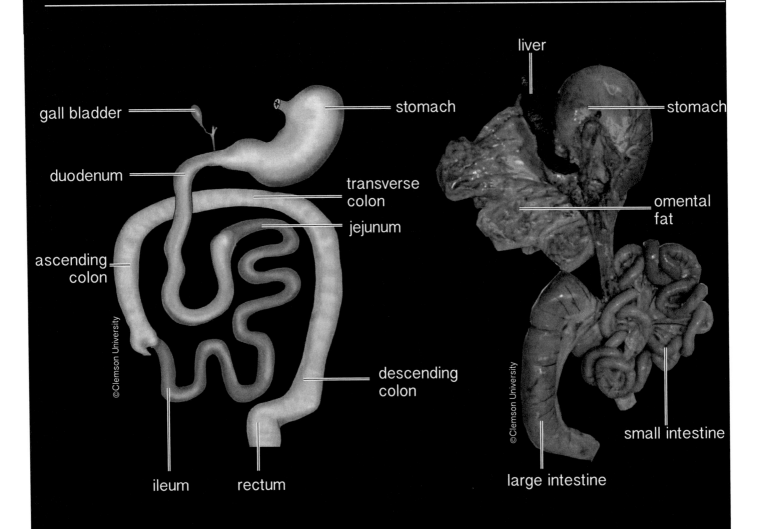

gall bladder

stomach

duodenum

transverse colon

jejunum

ascending colon

©Clemson University

descending colon

ileum

rectum

liver

stomach

omental fat

©Clemson University

small intestine

large intestine

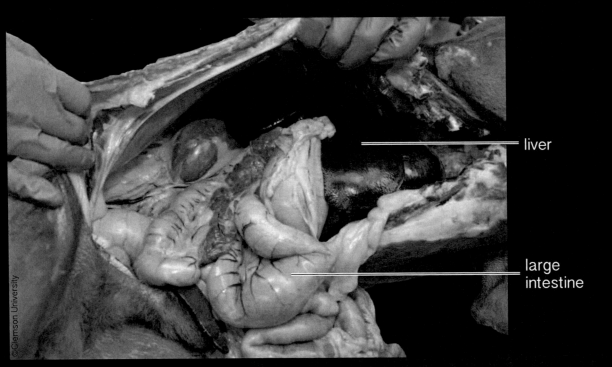

liver

large intestine

©Clemson University

Canine

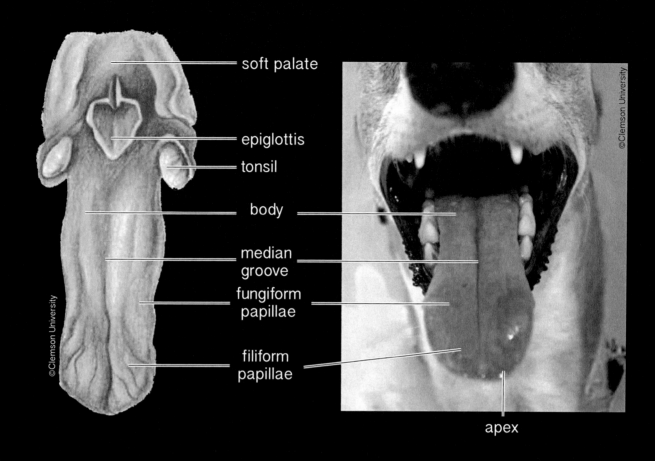

soft palate

epiglottis

tonsil

body

median groove

fungiform papillae

filiform papillae

apex

©Clemson University

©Clemson University

Feline

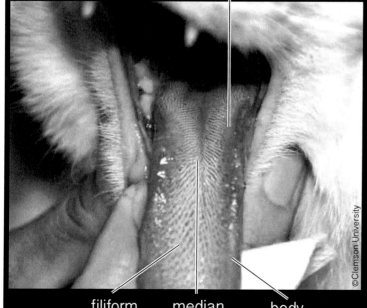

foliate papillae

filiform papillae

median groove

body

©Clemson University

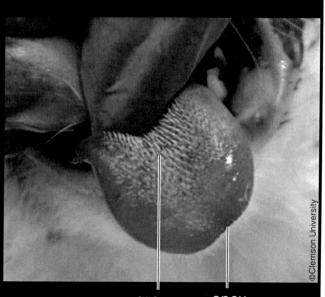

filiform papillae

apex

©Clemson University

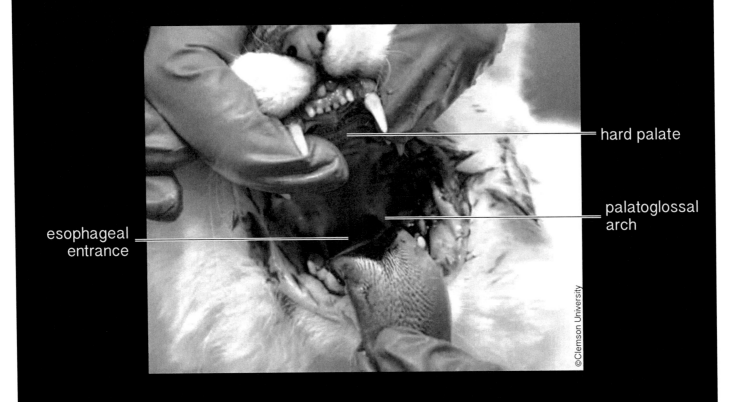

hard palate

palatoglossal arch

esophageal entrance

©Clemson University

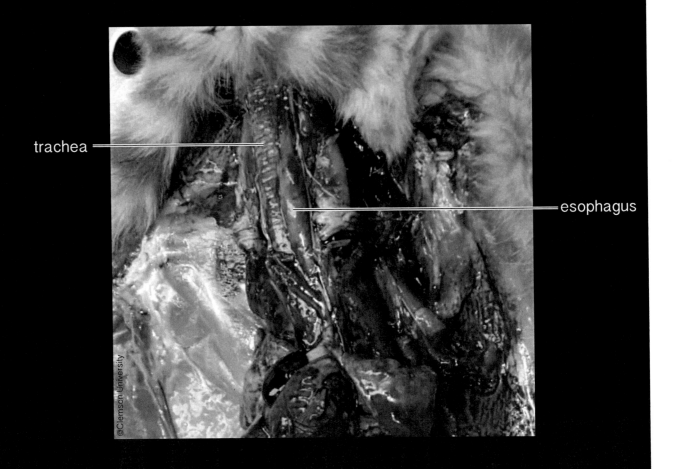

trachea

esophagus

©Clemson University

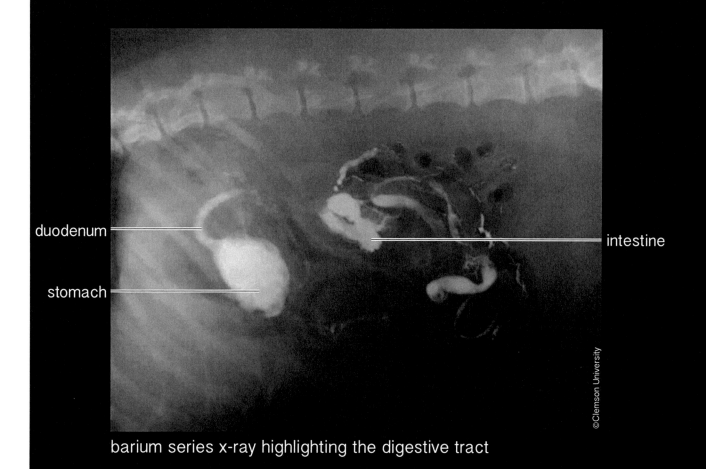

duodenum

stomach

intestine

©Clemson University

barium series x-ray highlighting the digestive tract

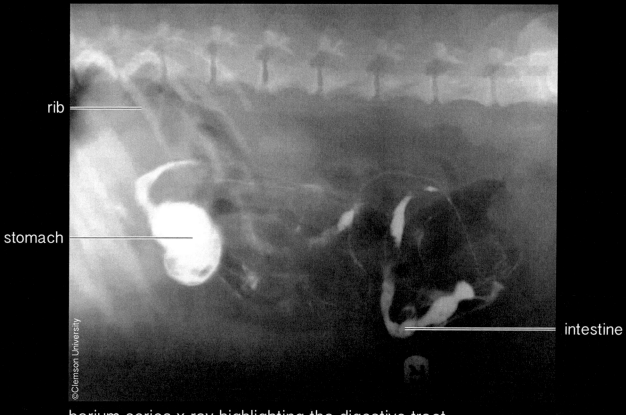

rib

stomach

intestine

©Clemson University

barium series x-ray highlighting the digestive tract

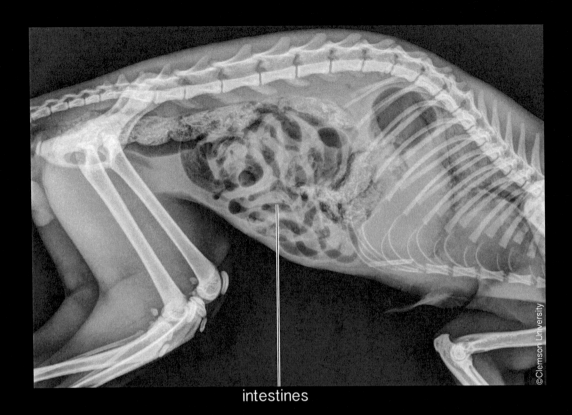

intestines

©Clemson University

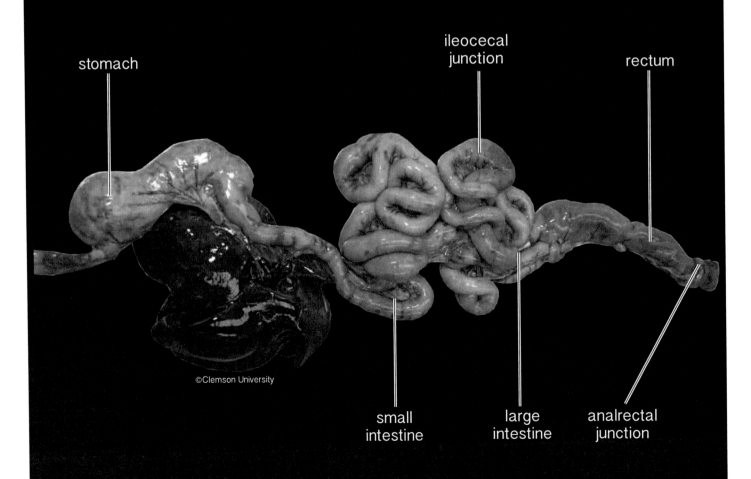

stomach

ileocecal
junction

rectum

©Clemson University

small
intestine

large
intestine

analrectal
junction

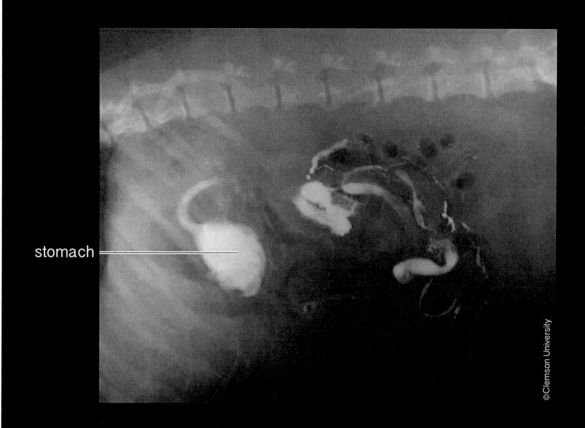

stomach

Feline

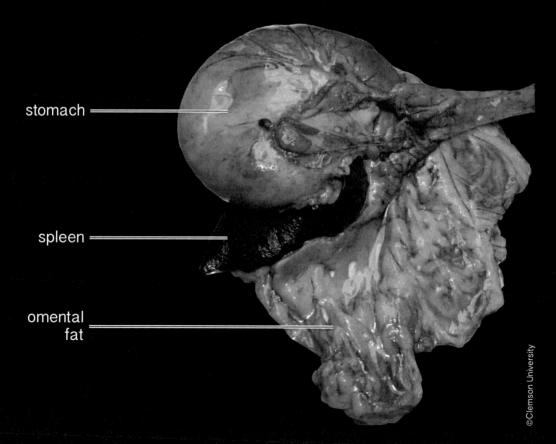

stomach

spleen

omental
fat

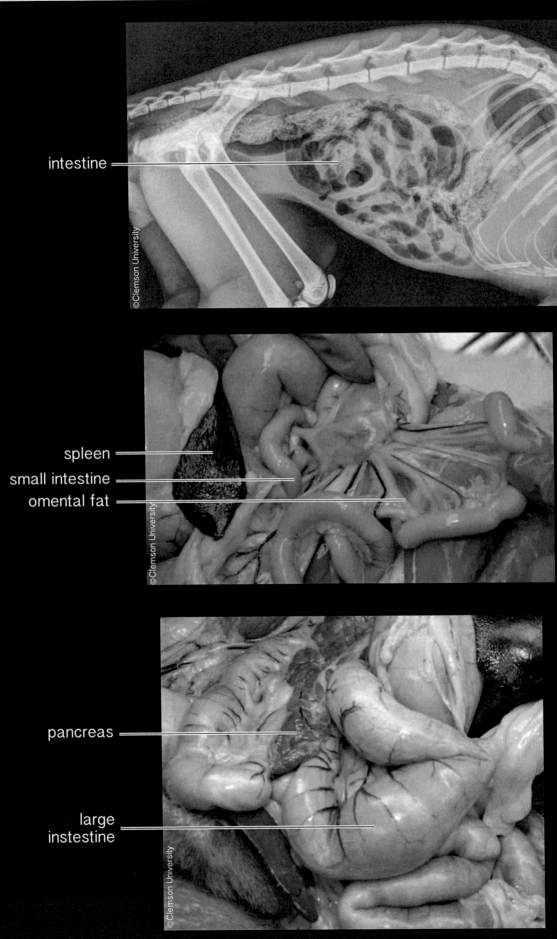

intestine

spleen
small intestine
omental fat

pancreas

large
instestine

©Clemson University

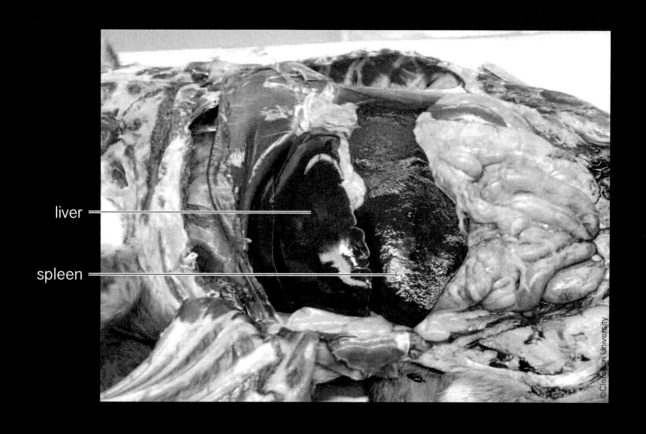

liver

spleen

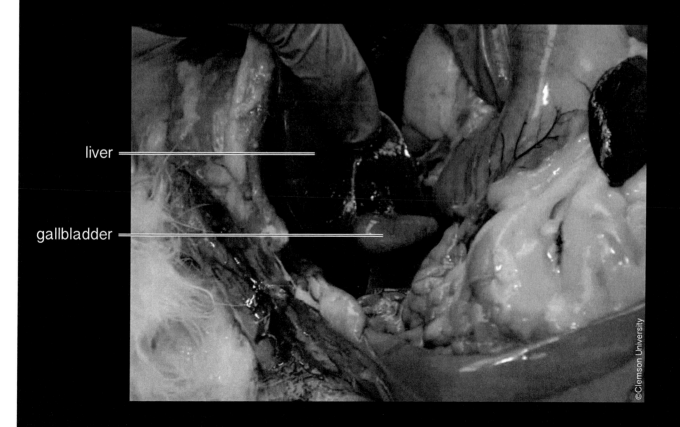

liver

gallbladder

Urinary System

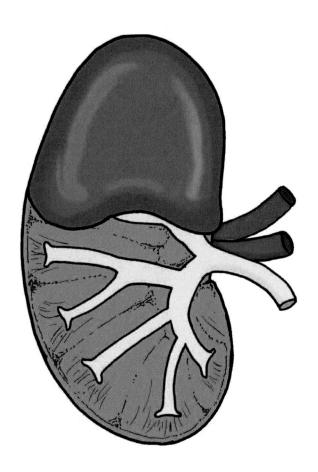 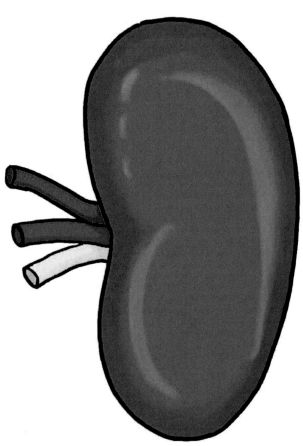

Kidney

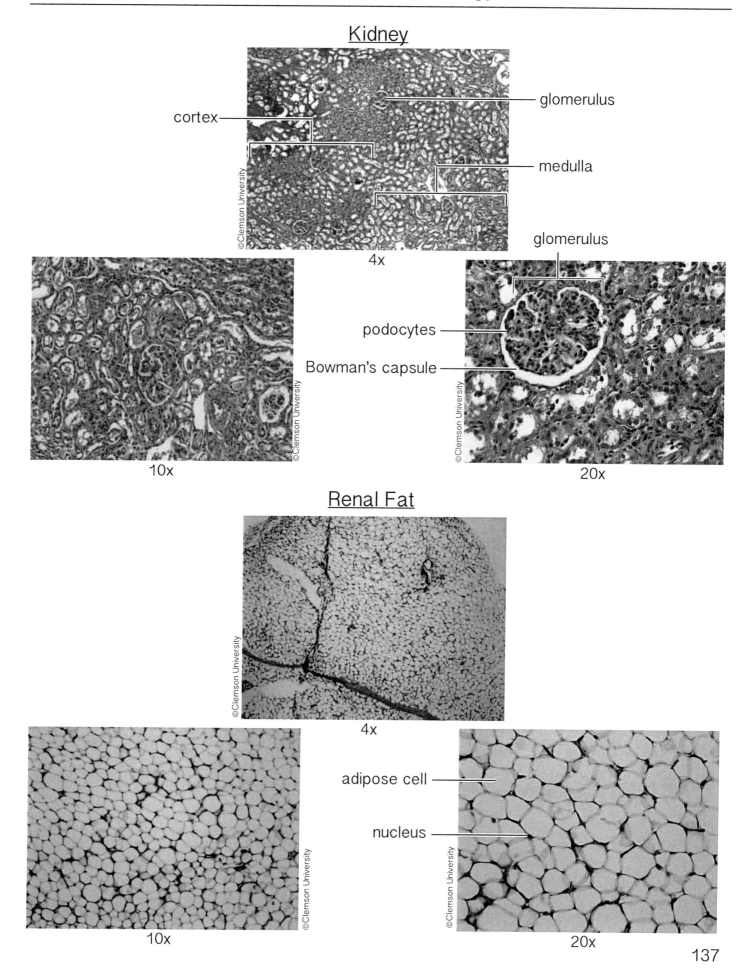

glomerulus

cortex

medulla

4x

glomerulus

podocytes

Bowman's capsule

10x

20x

Renal Fat

4x

adipose cell

nucleus

10x

20x

Histology - URINARY SYSTEM

Ureter

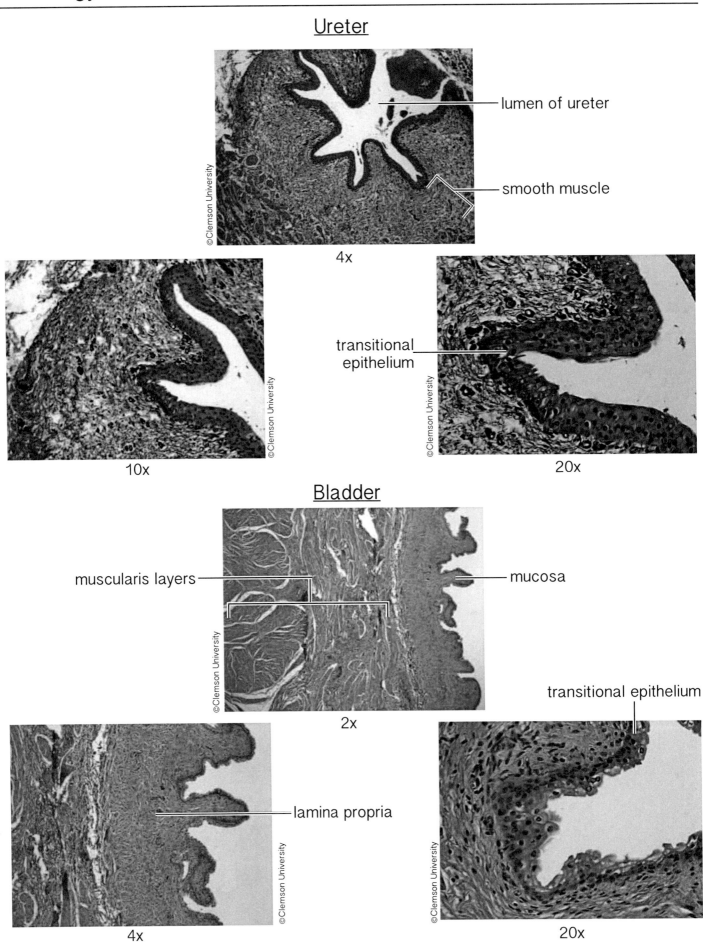

lumen of ureter

smooth muscle

4x

transitional epithelium

10x

20x

Bladder

muscularis layers

mucosa

2x

lamina propria

transitional epithelium

4x

20x

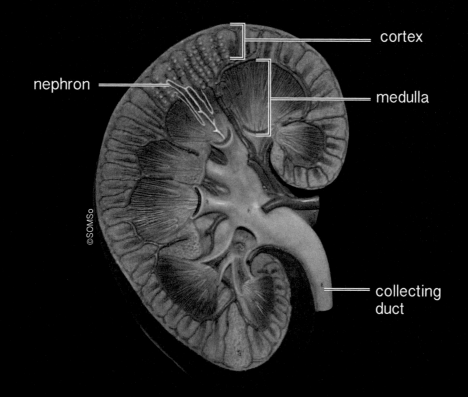

cortex

nephron

medulla

collecting
duct

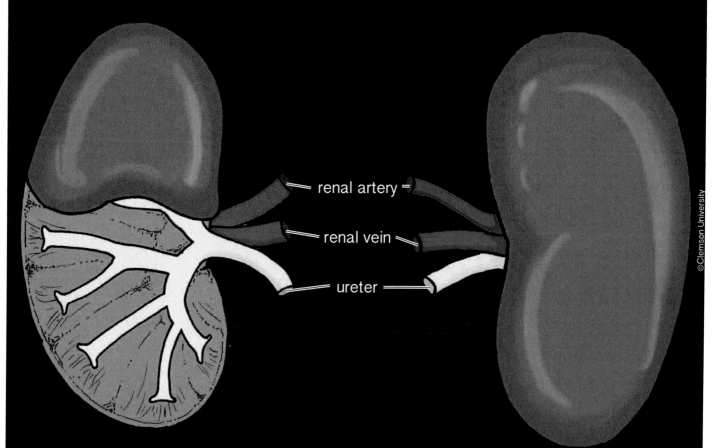

renal artery

renal vein

ureter

Glomerulus

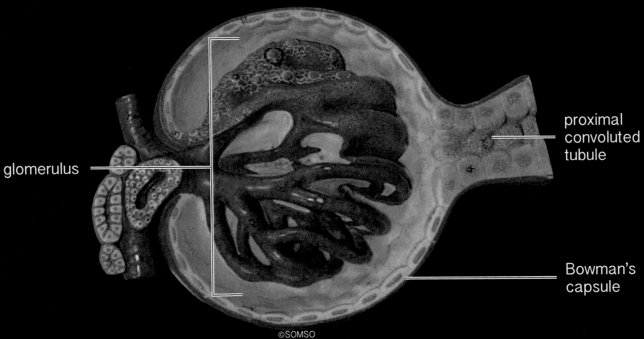

glomerulus

proximal convoluted tubule

Bowman's capsule

©SOMSO

Nephron

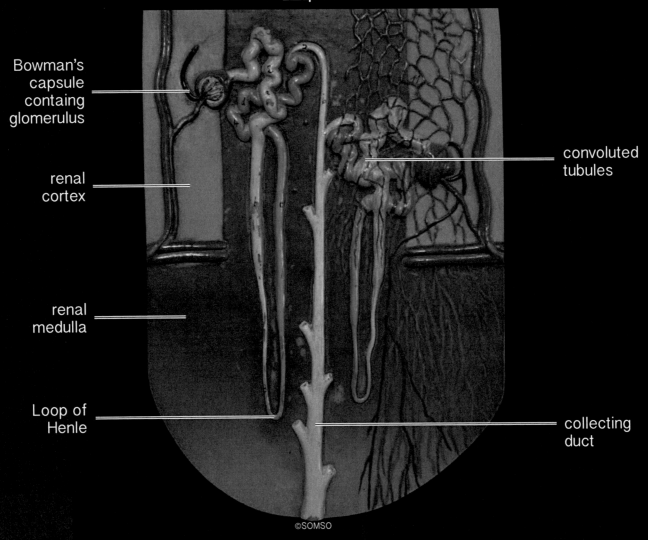

Bowman's capsule containg glomerulus

renal cortex

renal medulla

Loop of Henle

convoluted tubules

collecting duct

©SOMSO

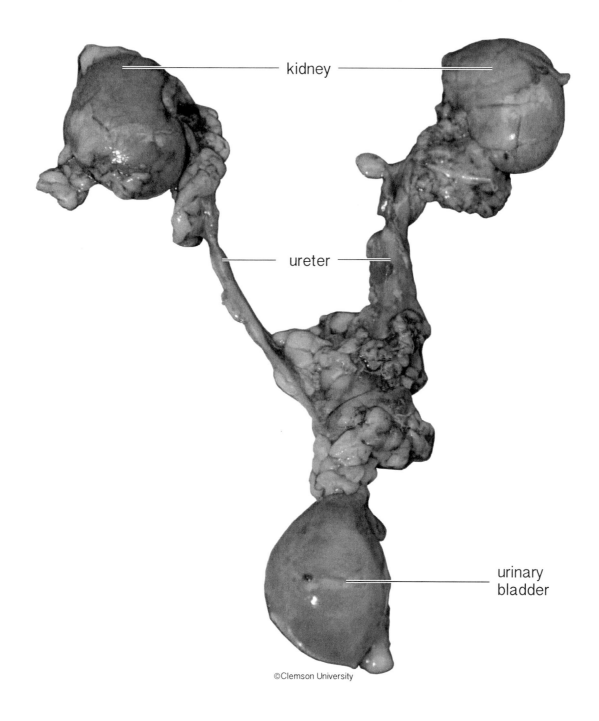

kidney

ureter

urinary
bladder

©Clemson University

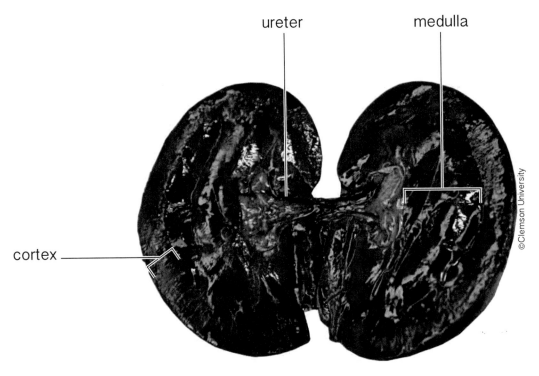

ureter   medulla

cortex

mid-saggital section of kidney

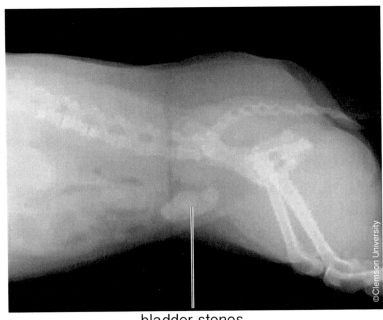

bladder stones

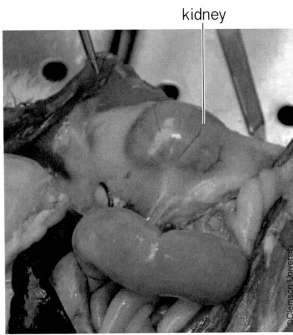

kidney

kidney in renal fat

Endocrine/Reproductive Systems

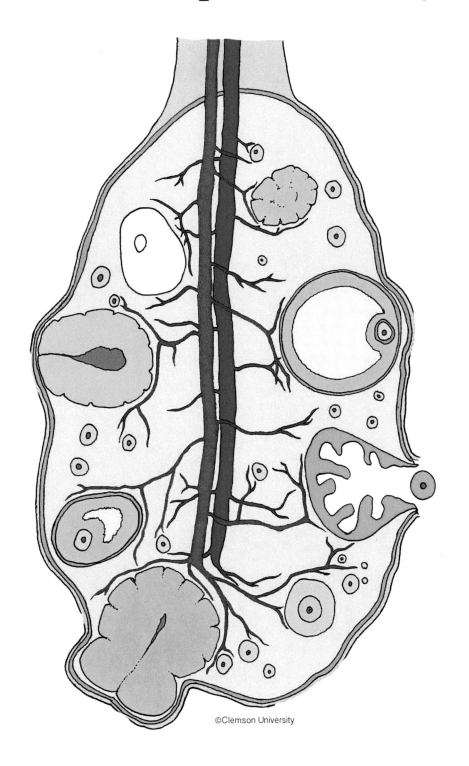

Histology - ENDOCRINE SYSTEM

Adrenal Gland

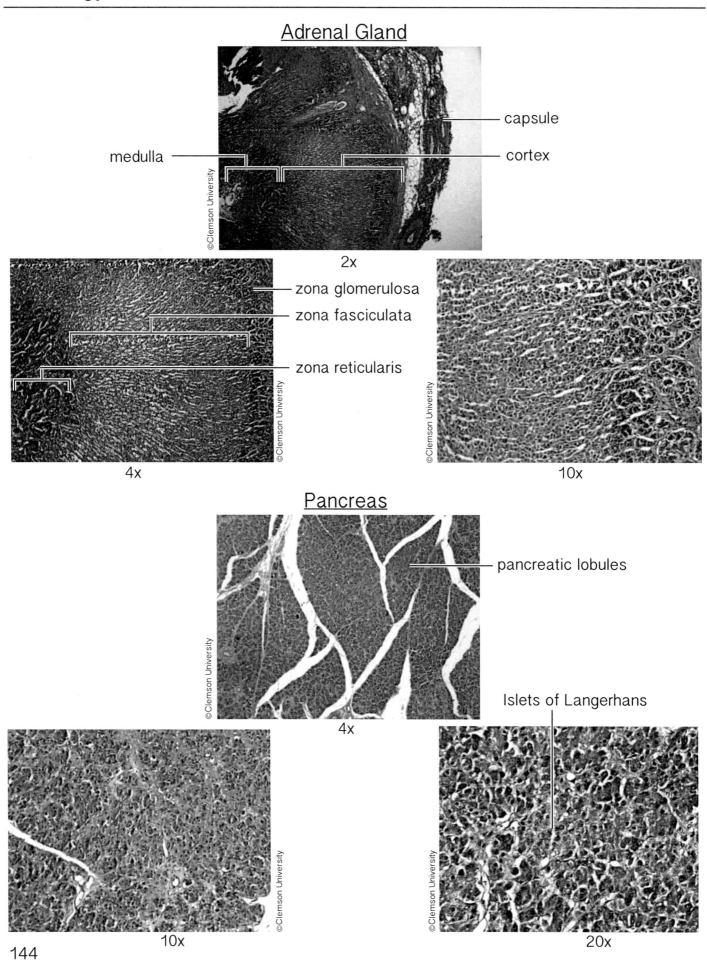

capsule

medulla

cortex

2x

zona glomerulosa

zona fasciculata

zona reticularis

4x

10x

Pancreas

pancreatic lobules

4x

Islets of Langerhans

10x

20x

Histology - ENDOCRINE SYSTEM

Pineal Gland

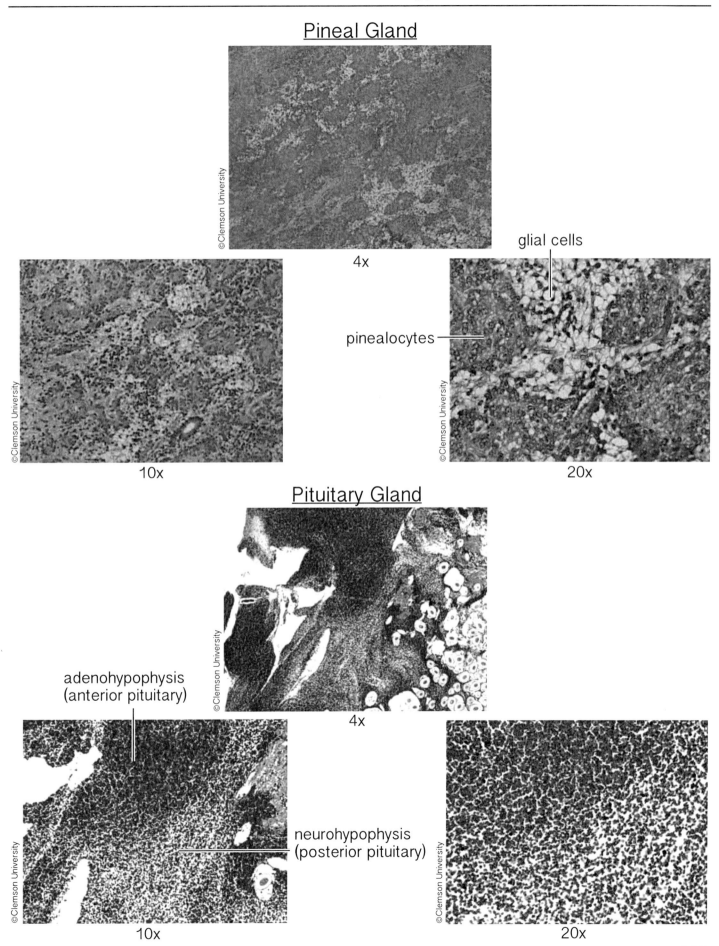

4x

10x

glial cells

pinealocytes

20x

Pituitary Gland

4x

adenohypophysis
(anterior pituitary)

neurohypophysis
(posterior pituitary)

10x

20x

145

Histology - ENDOCRINE SYSTEM

Thyroid

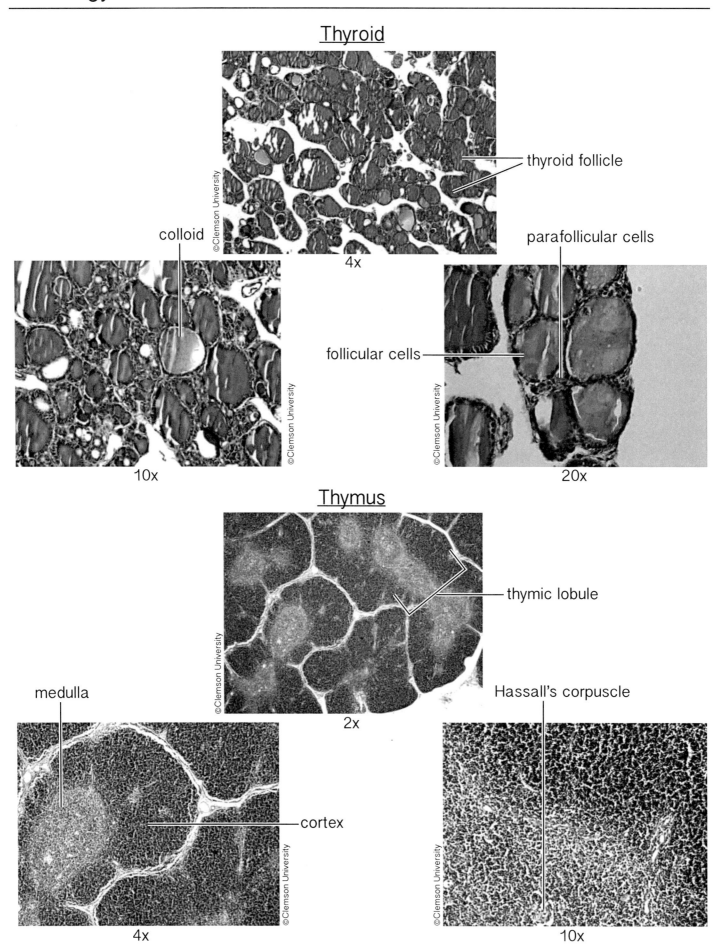

thyroid follicle

colloid

parafollicular cells

4x

follicular cells

10x

20x

Thymus

thymic lobule

2x

medulla

Hassall's corpuscle

cortex

4x

10x

Cervix

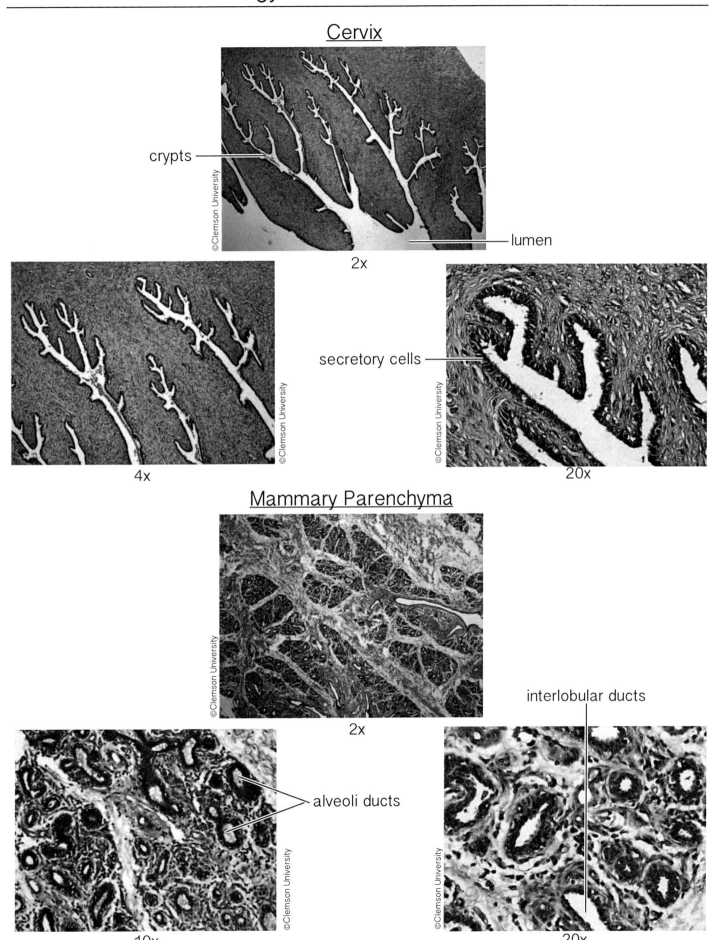

crypts

lumen

2x

4x

secretory cells

20x

Mammary Parenchyma

2x

10x

alveoli ducts

interlobular ducts

20x

Histology - FEMALE REPRODUCTIVE SYSTEM

Ovary

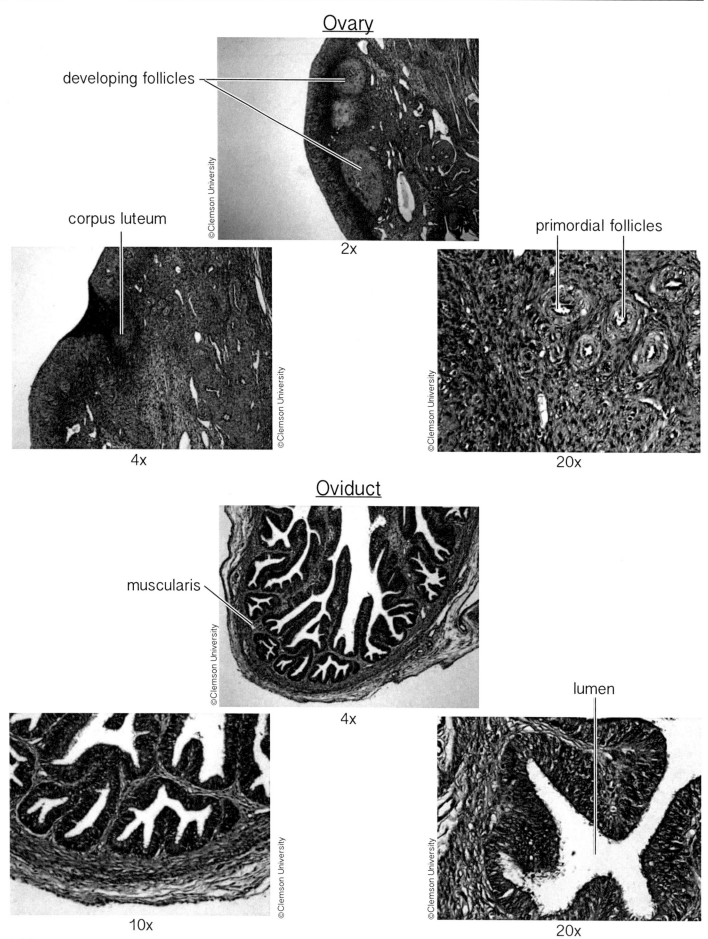

developing follicles

2x

corpus luteum

4x

primordial follicles

20x

Oviduct

muscularis

4x

10x

lumen

20x

©Clemson University

Teat

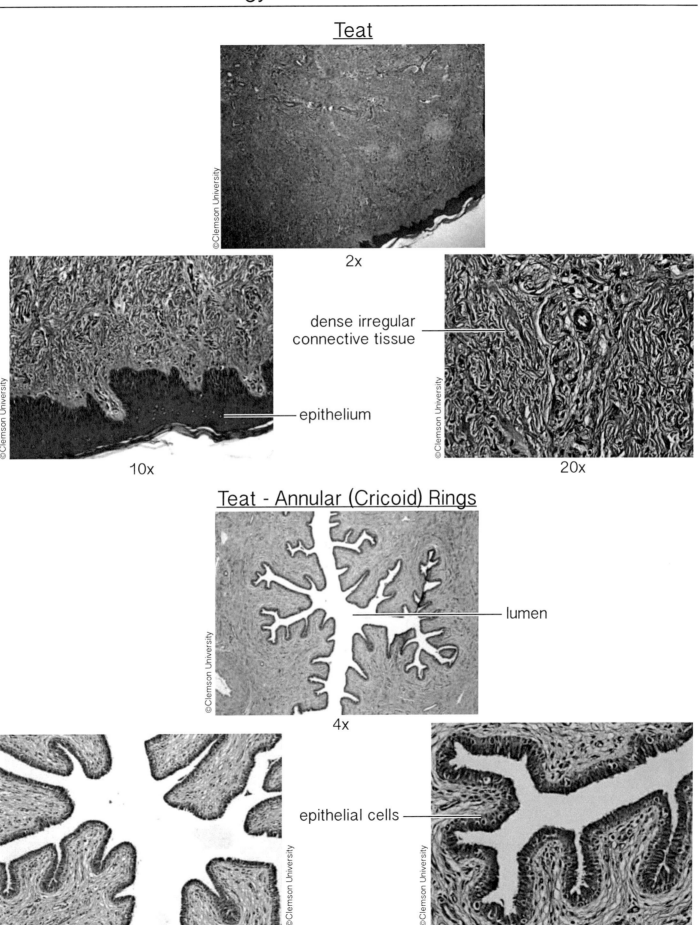

2x

dense irregular
connective tissue

epithelium

10x

20x

Teat - Annular (Cricoid) Rings

lumen

4x

epithelial cells

10x

20x

Histology - FEMALE REPRODUCTIVE SYSTEM

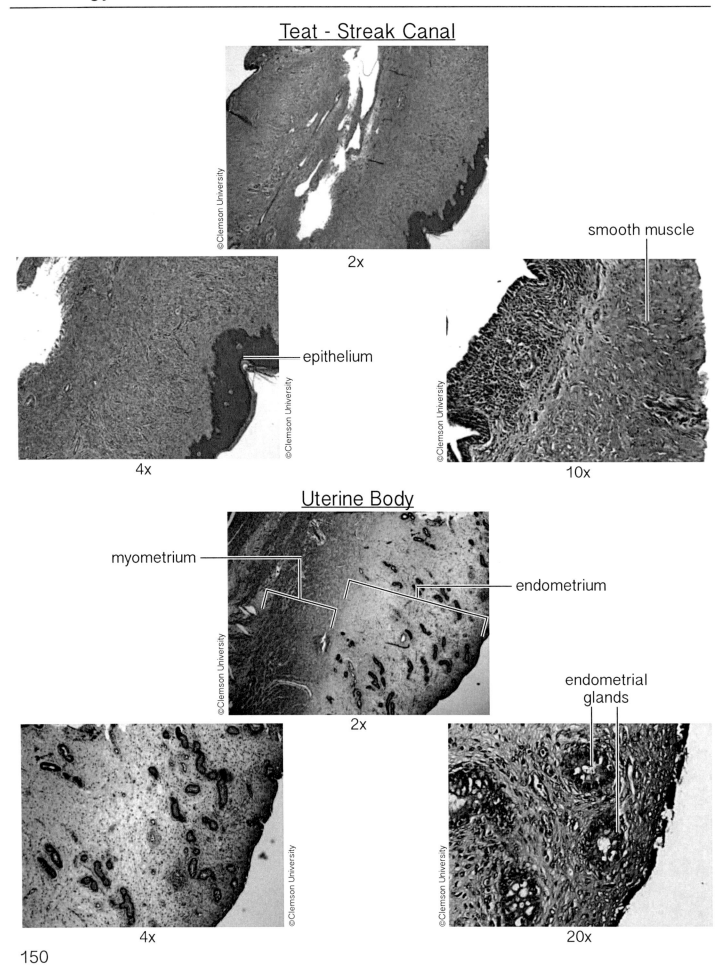

Teat - Streak Canal

2x

4x — epithelium

smooth muscle

10x

Uterine Body

myometrium — endometrium

2x

4x

endometrial glands

20x

Uterine Horn

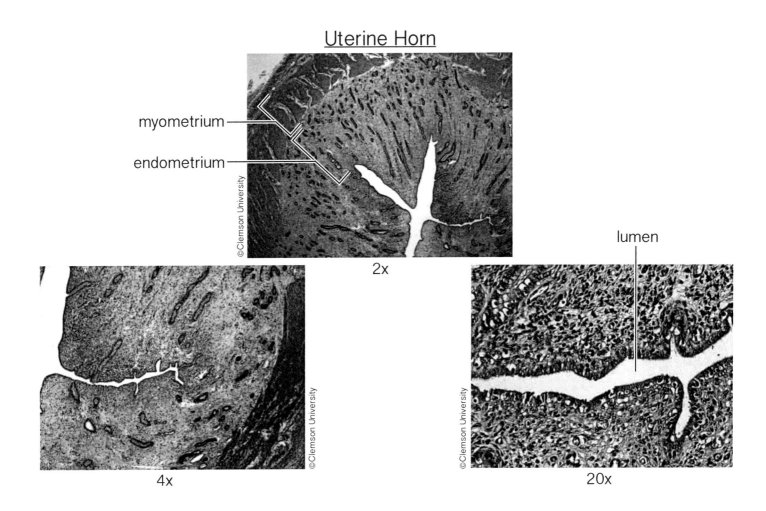

myometrium

endometrium

©Clemson University

2x

4x

©Clemson University

lumen

©Clemson University

20x

Histology - MALE REPRODUCTIVE SYSTEM

Head of Epididymis

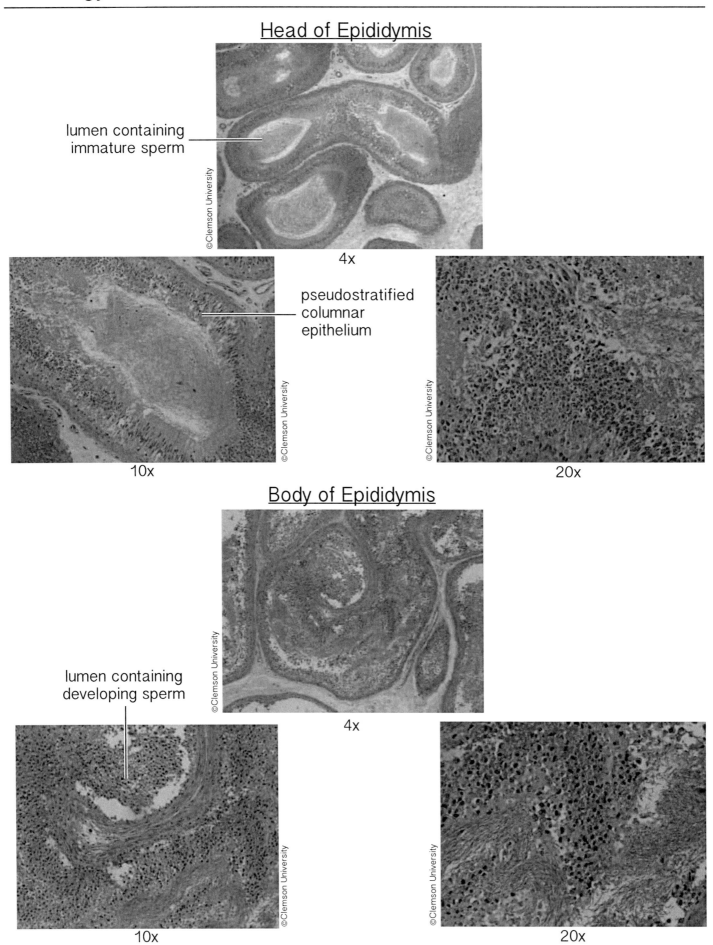

lumen containing
immature sperm

©Clemson University

4x

pseudostratified
columnar
epithelium

©Clemson University

10x

©Clemson University

20x

Body of Epididymis

©Clemson University

4x

lumen containing
developing sperm

©Clemson University

10x

©Clemson University

20x

Tail of Epididymis

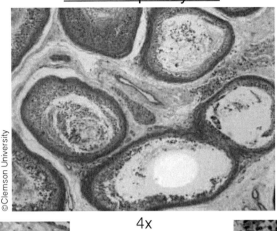

4x

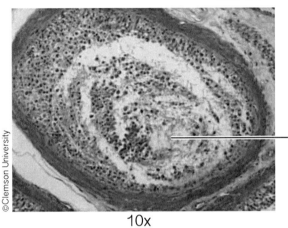

10x

lumen containing
mature sperm

smooth muscle

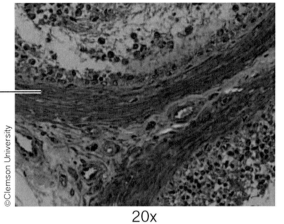

20x

Spermatic Cord

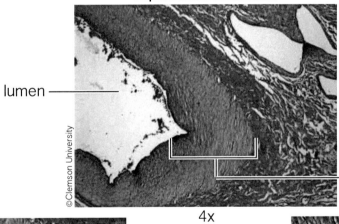

lumen

smooth muscle

4x

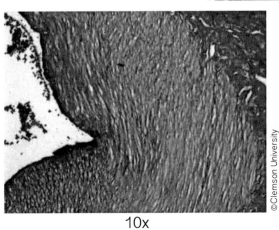

10x

epithelium

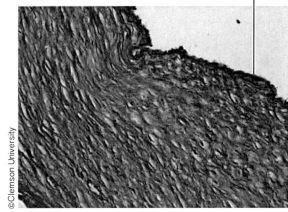

20x

Histology - MALE REPRODUCTIVE SYSTEM

Testis

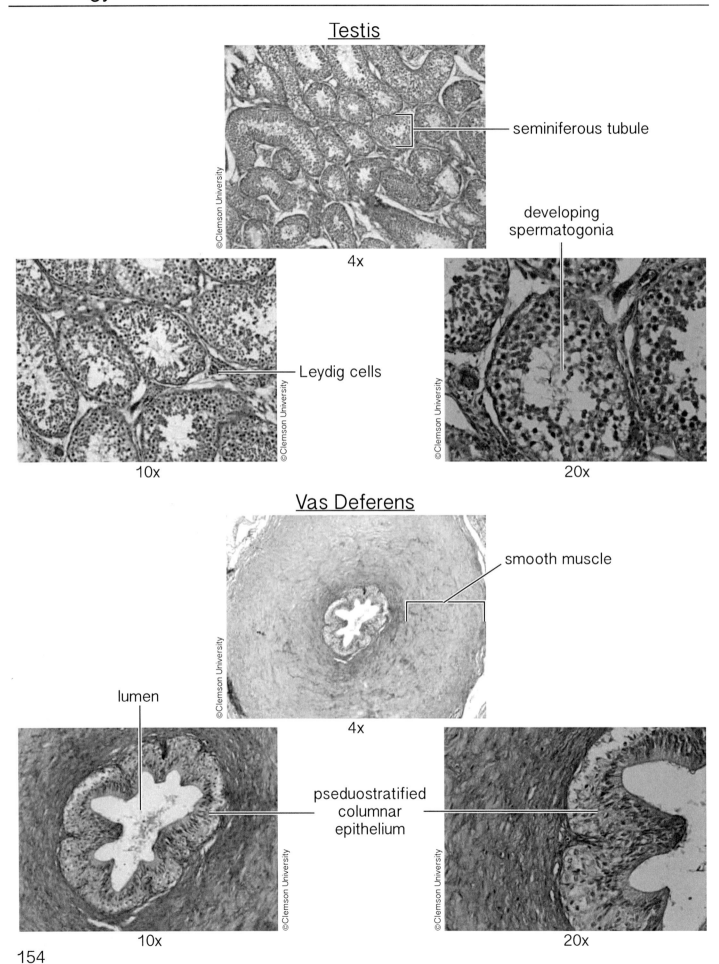

seminiferous tubule

4x

developing spermatogonia

Leydig cells

10x

20x

Vas Deferens

smooth muscle

4x

lumen

pseduostratified columnar epithelium

10x

20x

Canine

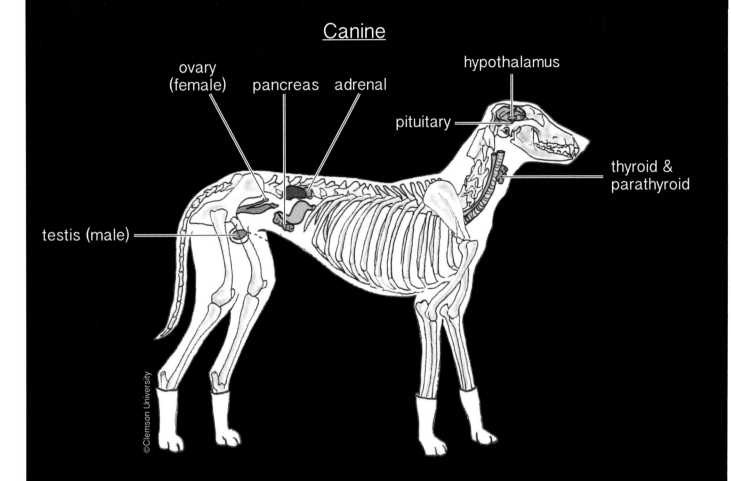

ovary (female)   pancreas   adrenal   hypothalamus   pituitary   thyroid & parathyroid   testis (male)

Feline

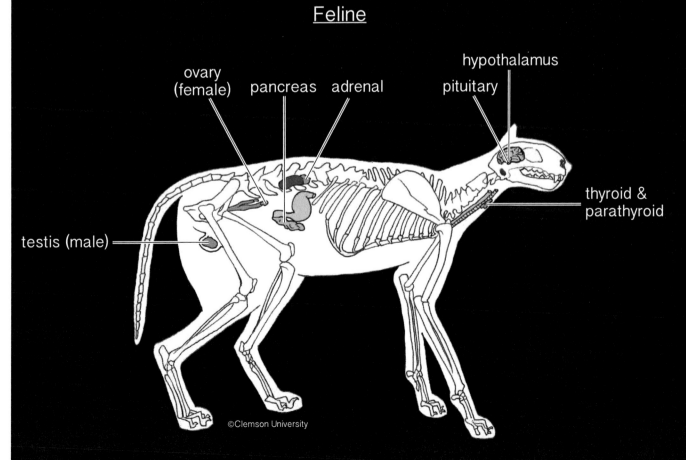

ovary (female)   pancreas   adrenal   hypothalamus   pituitary   thyroid & parathyroid   testis (male)

Endocrine System - PANCREAS

Canine

pancreas

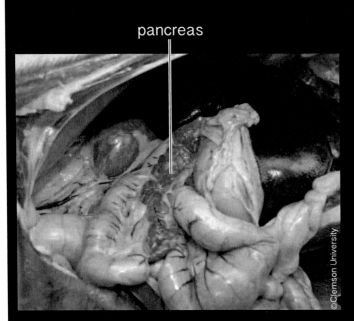

pancreas

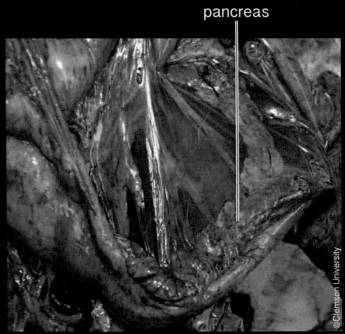

Feline

omental fat

pancreas

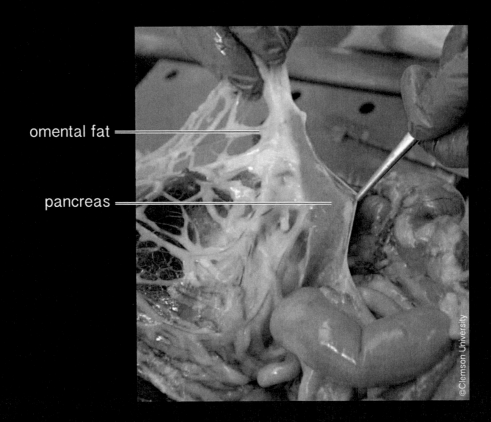

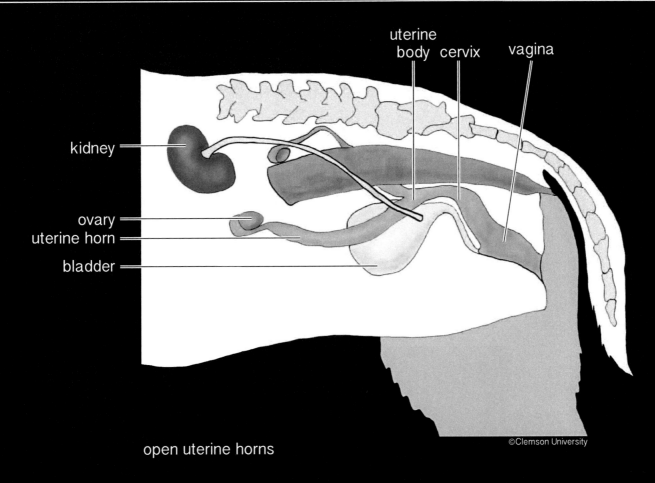

open uterine horns

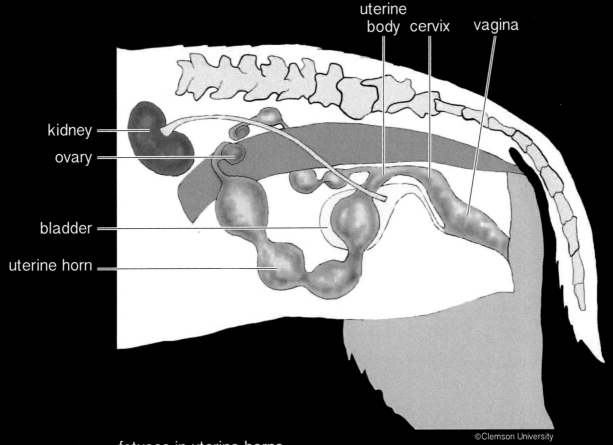

fetuses in uterine horns

Female Reproductive System - EXTERNAL GENITALIA

Canine

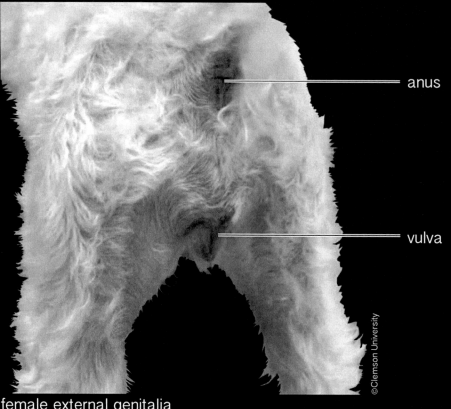

anus

vulva

©Clemson University

female external genitalia

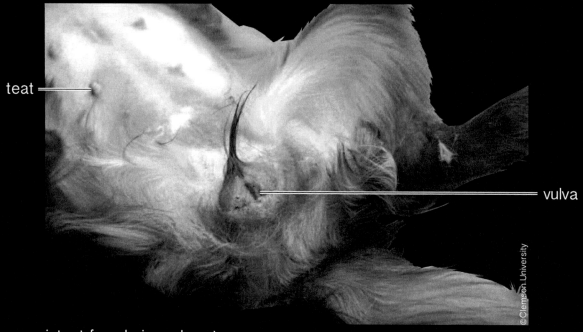

teat

vulva

©Clemson University

intact female in early estrus

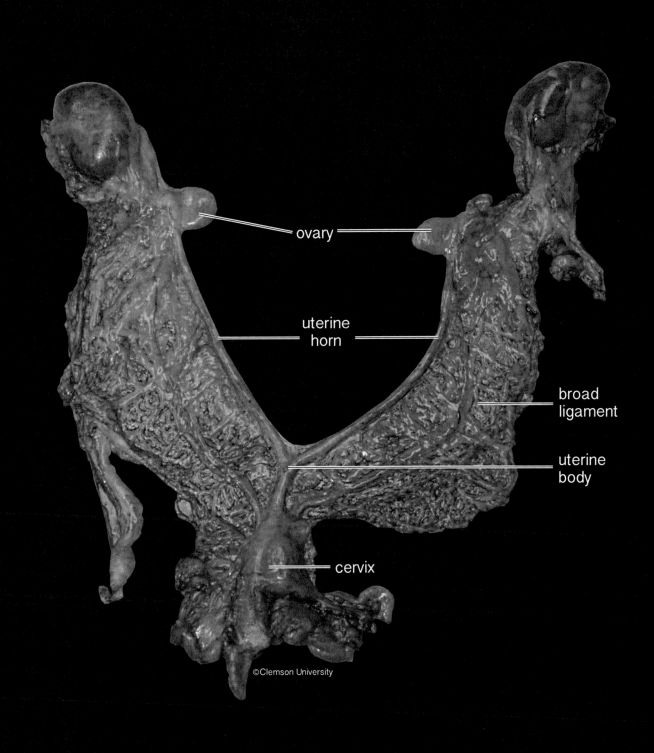

ovary

uterine
horn

broad
ligament

uterine
body

cervix

©Clemson University

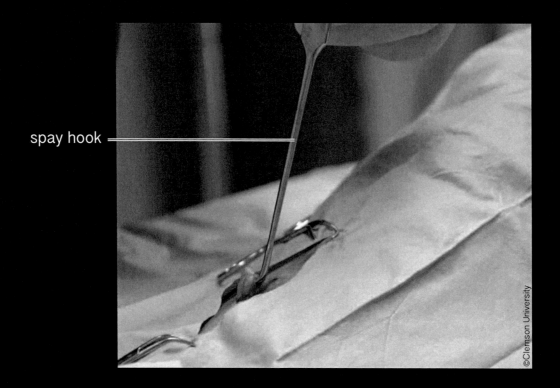

spay hook

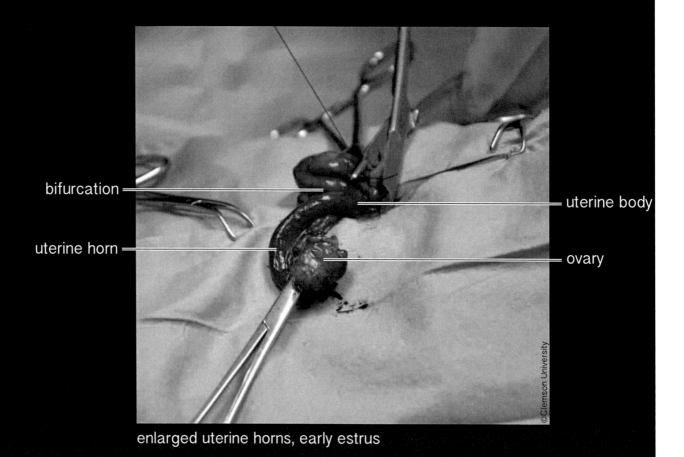

bifurcation

uterine horn

uterine body

ovary

enlarged uterine horns, early estrus

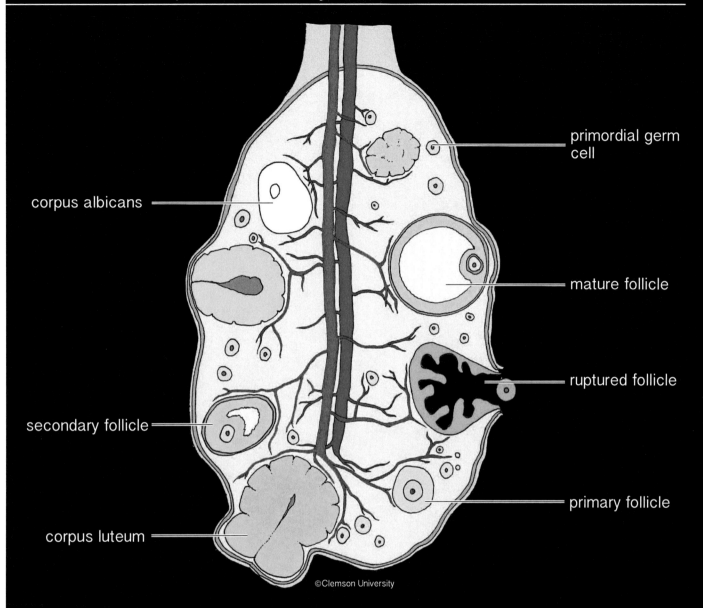

primordial germ cell

corpus albicans

mature follicle

ruptured follicle

secondary follicle

primary follicle

corpus luteum

©Clemson University

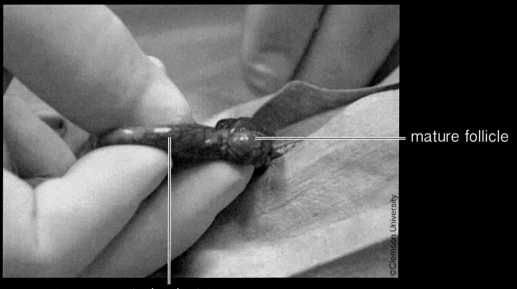

mature follicle

uterine horn

©Clemson University

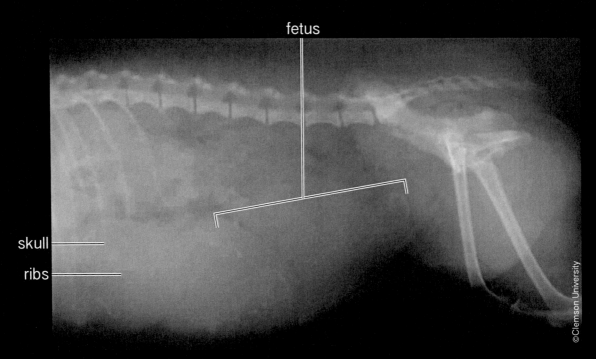

Puppies in utero

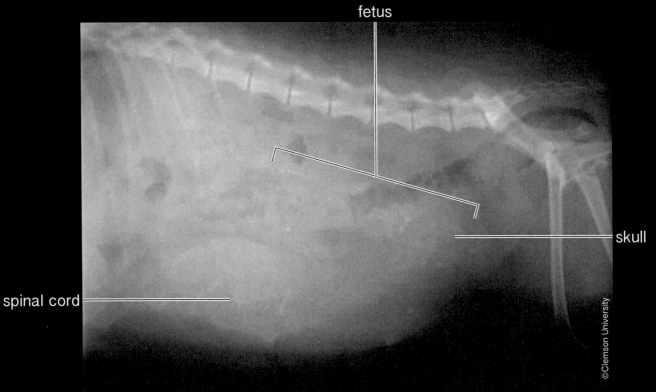

Five puppies in utero, late gestation

Canine

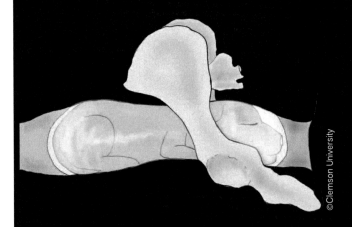

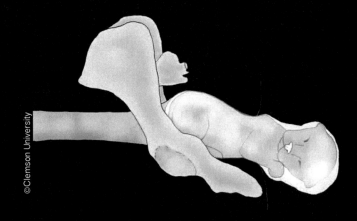

early parturition

late parturition

Feline

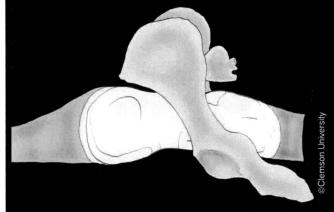

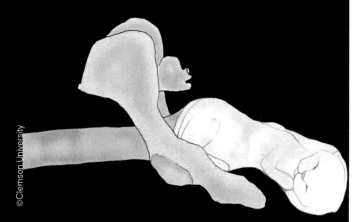

early parturition

late parturition

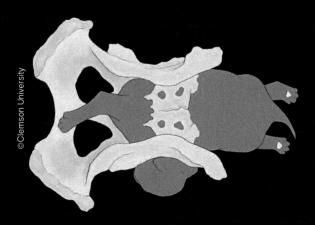

dystocia

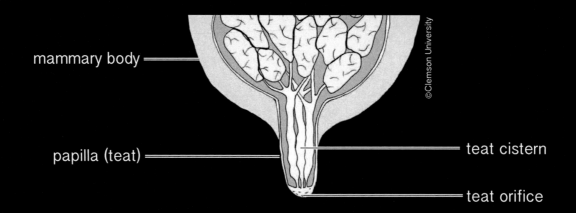

mammary body

papilla (teat)

teat cistern

teat orifice

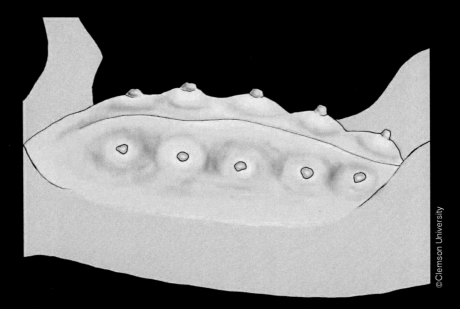

mammary gland of pregnant canine

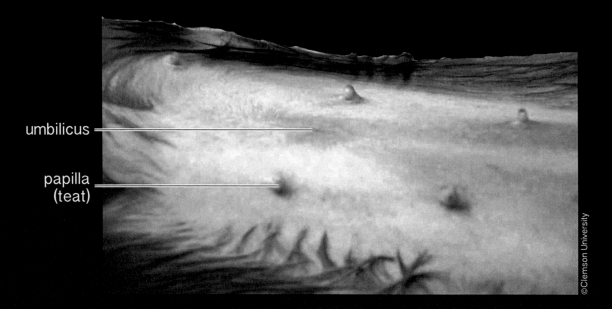

umbilicus

papilla
(teat)

Canine

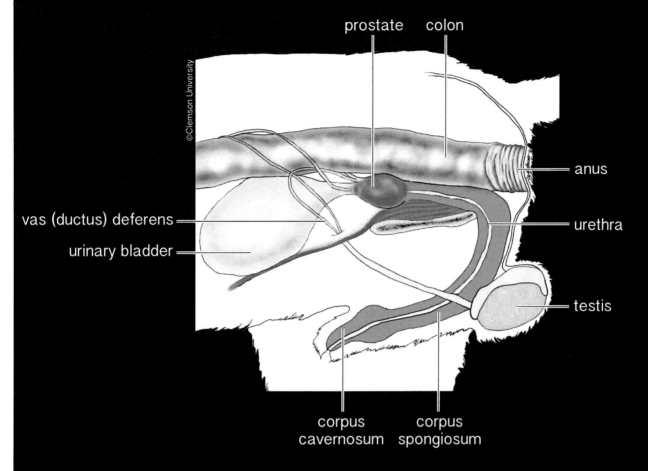

prostate

colon

©Clemson University

anus

vas (ductus) deferens

urethra

urinary bladder

testis

corpus
cavernosum

corpus
spongiosum

Feline

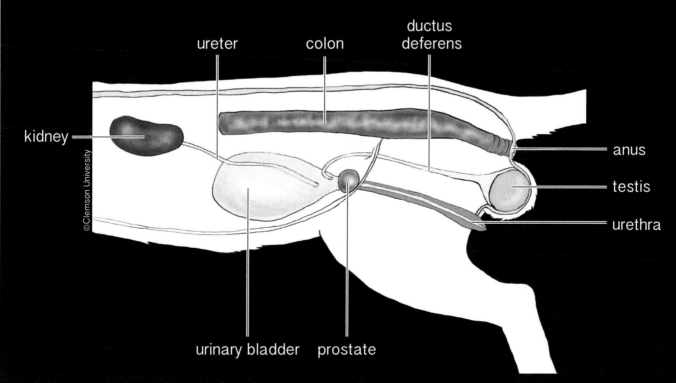

ureter

colon

ductus
deferens

kidney

©Clemson University

anus

testis

urethra

urinary bladder

prostate

MALE REPRODUCTIVE SYSTEM

Canine

penis

testis in scrotal sac

spermatic cord

tunica albuginea

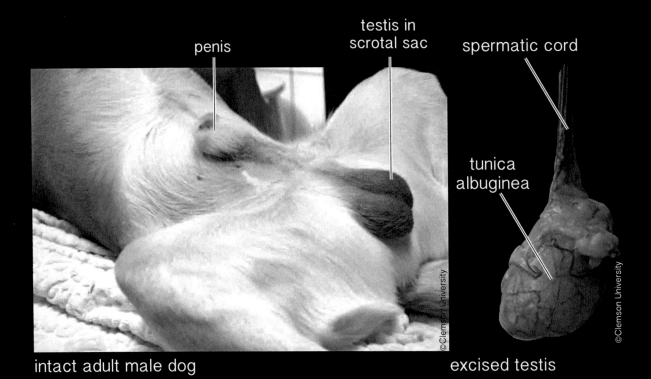

intact adult male dog

excised testis

Feline

testis in scrotal sac

penis

spermatic cord

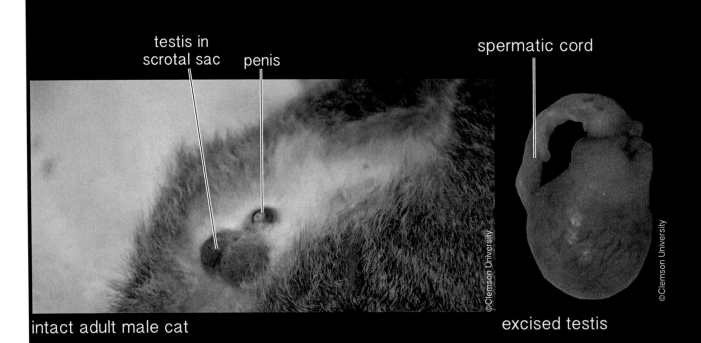

intact adult male cat

excised testis

Canine

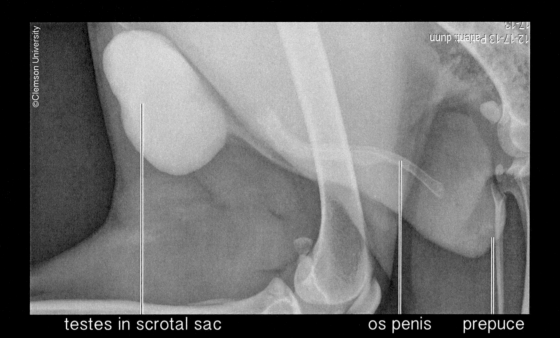

testes in scrotal sac   os penis   prepuce

Feline

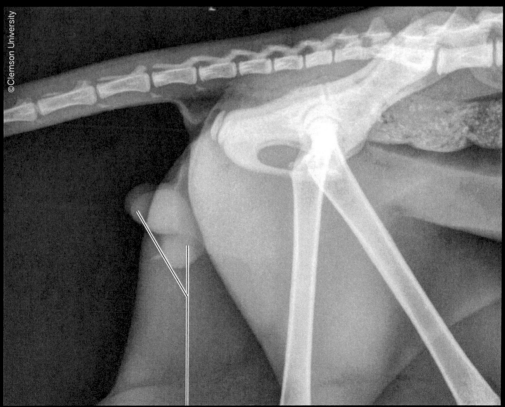

testes in scrotal sac

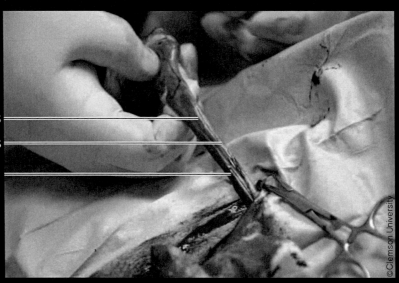

pampiniform plexus
ductus deferens
cremaster muscle

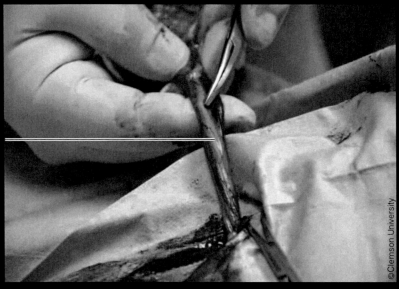

ductus deferens

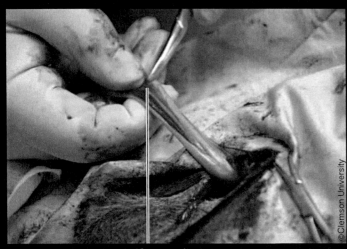

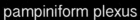

pampiniform plexus

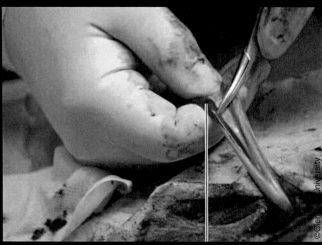

cremaster muscle

Chapter 11

Veterinary Techniques

©Clemson University

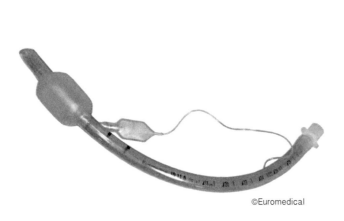

©Euromedical

Intubation tube

©Accu-Scope

Microscope

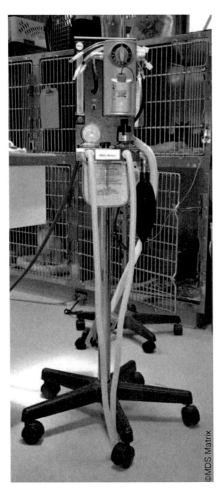

©MDS Matrix

Mobile anesthesia machine

©Baxter

Portable IV infusion pump
on an IV pole

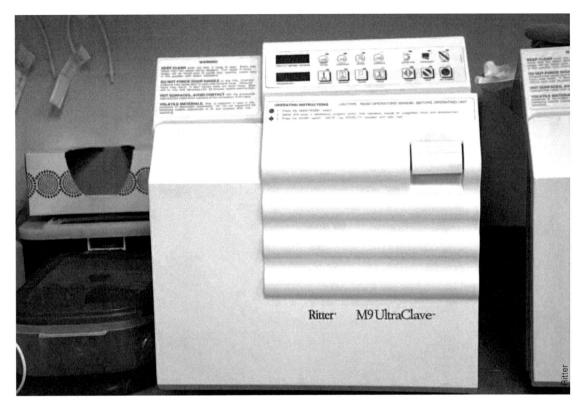

Autoclave

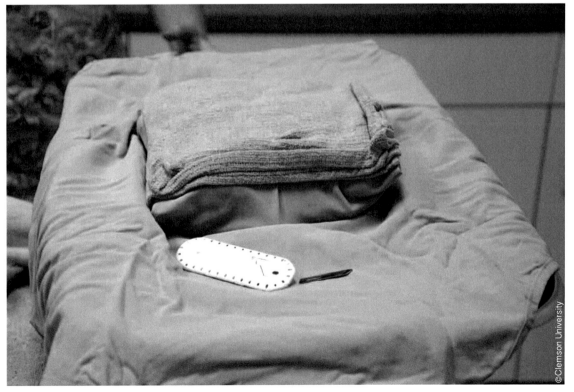

Surgical pack

Surgical instruments

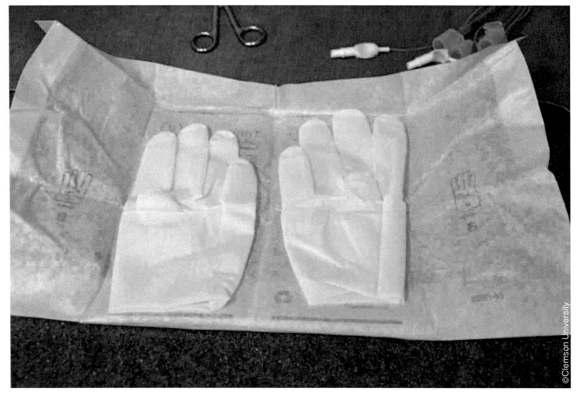

Sterile surgical gloves

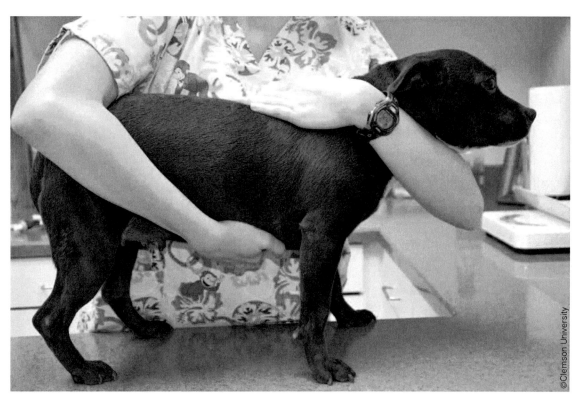

Standing position

Lateral recumbency

Open-ended cloth muzzle, profile

Open-ended cloth muzzle, front

Cloth cat mask

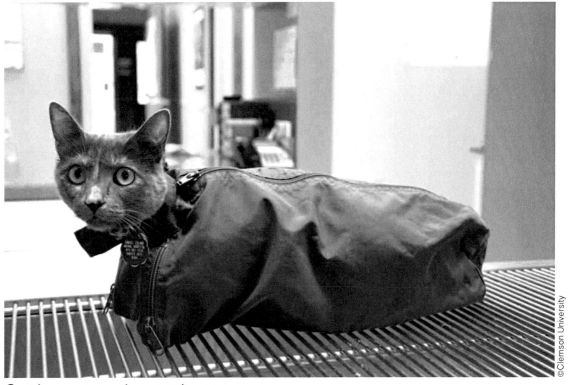

Cat bag, sternal recumbency

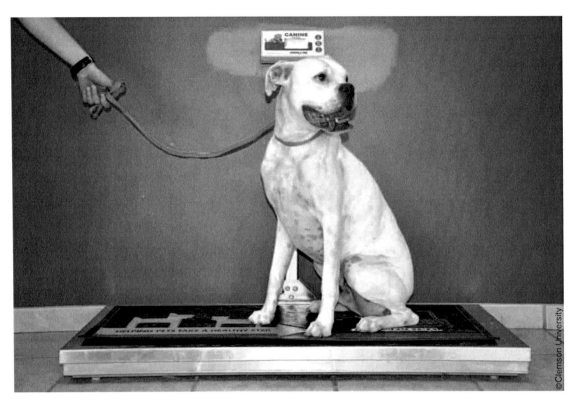

Mid-size floor scale

Small animal scale

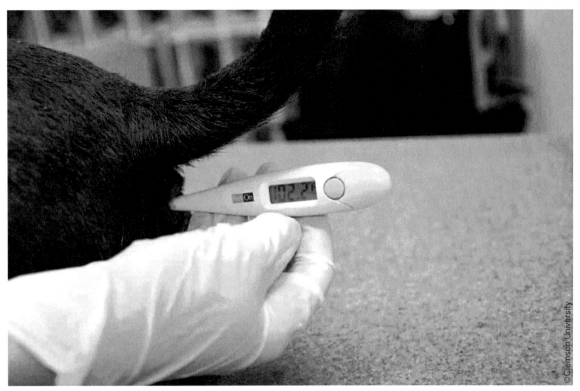

Rectal temperature

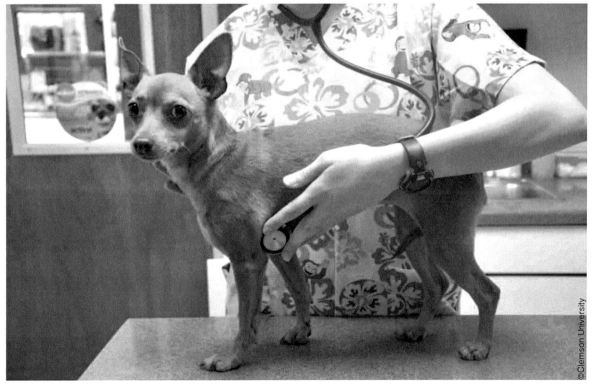

Determining heart rate and respiration with a stethoscope

Nail trimming

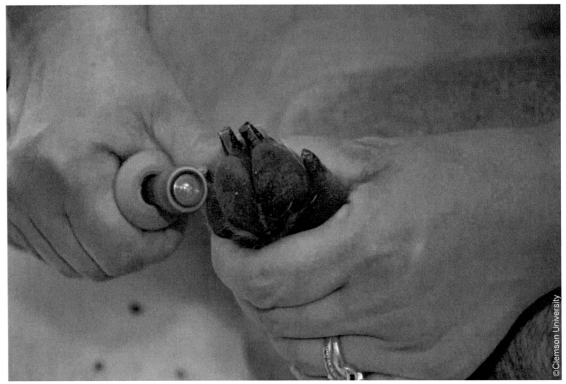

Nail filing

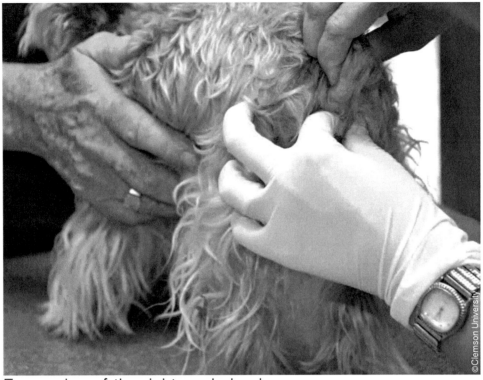

Expression of the right anal gland

Expression of the left anal gland

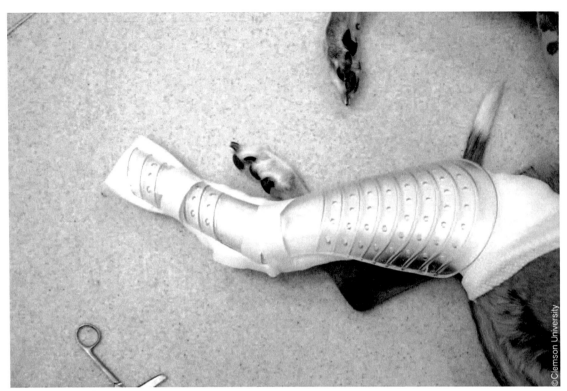

Unwrapped cast

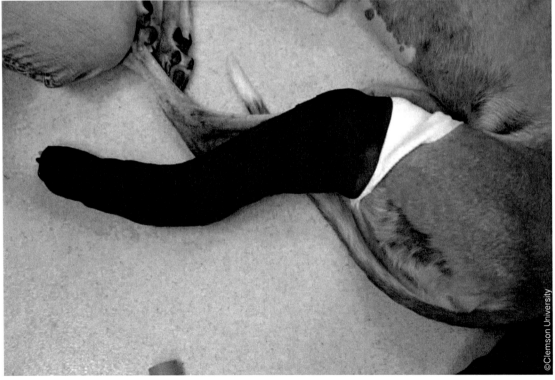

Wrapped cast

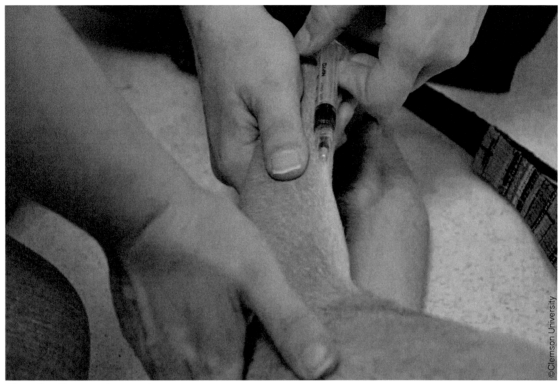

Cephalic venipuncture

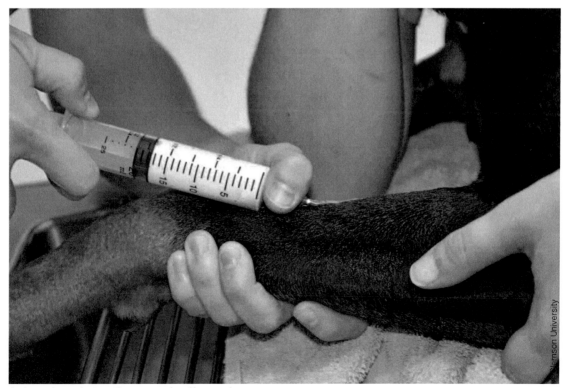

Injection into the cephalic vein

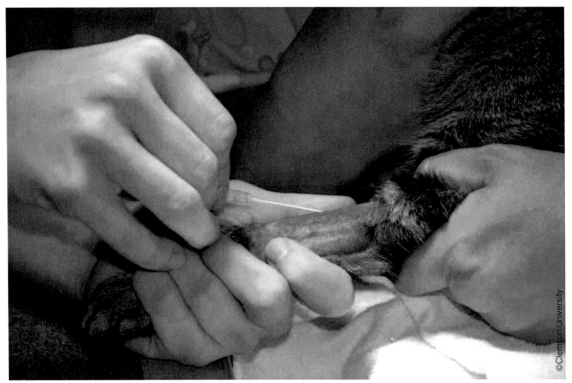

Insertion of a catheter into the cephalic vein

Capped catheter

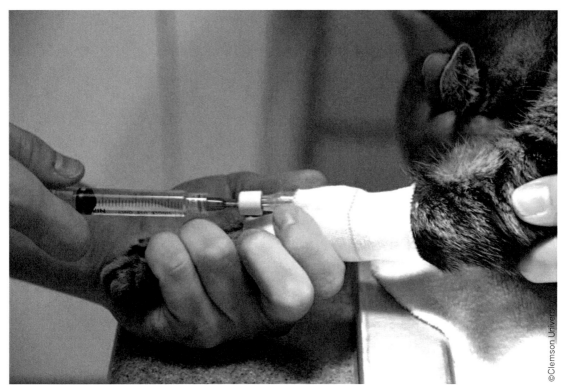

Flushing a catheter

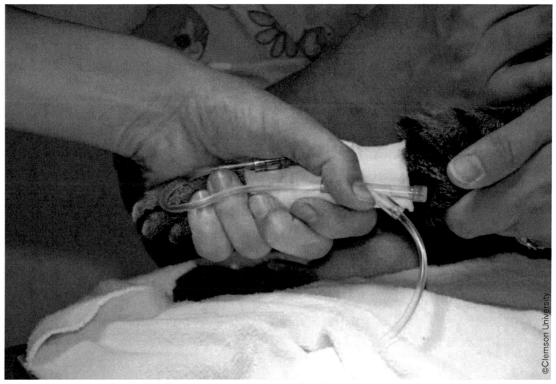

Attachment of an IV fluids line to a cathether

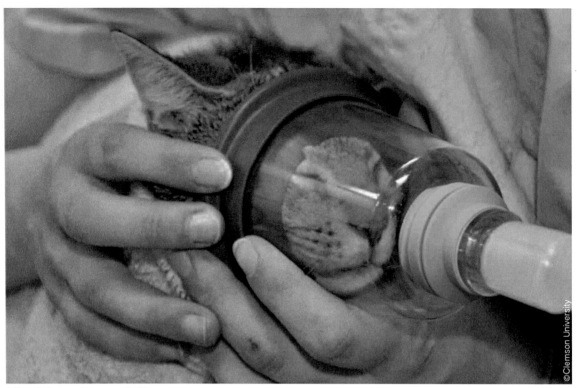

Masking down, feline

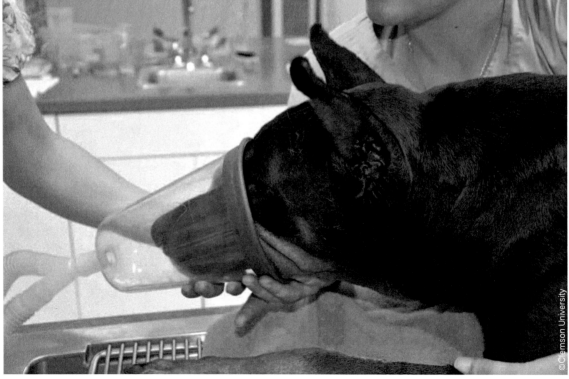

Masking down, canine

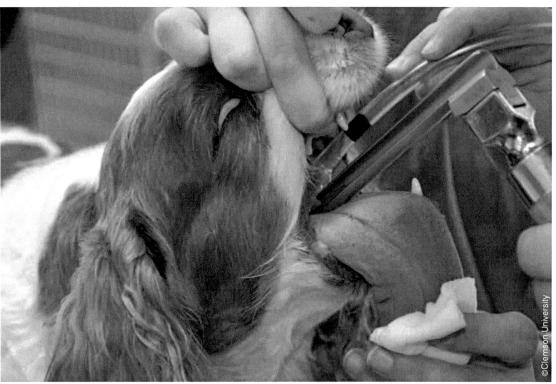

Intubation with a laryngoscope

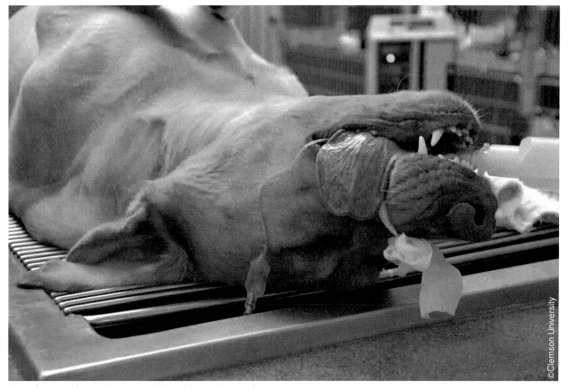

Intubated and anesthesized patient

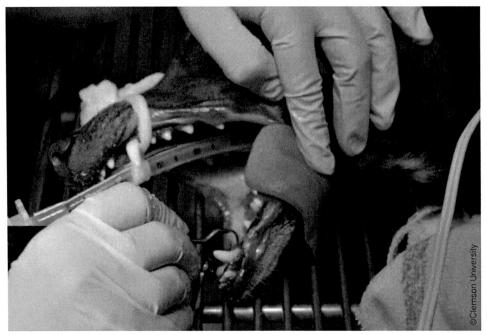

Dental surgery

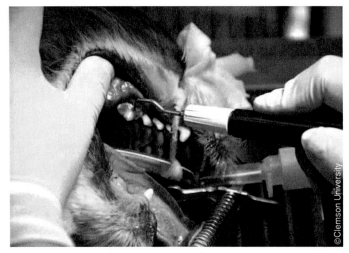

Cleaning with subgingival curette

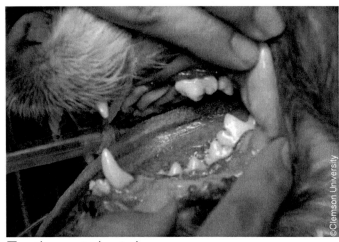

Teeth post-dental

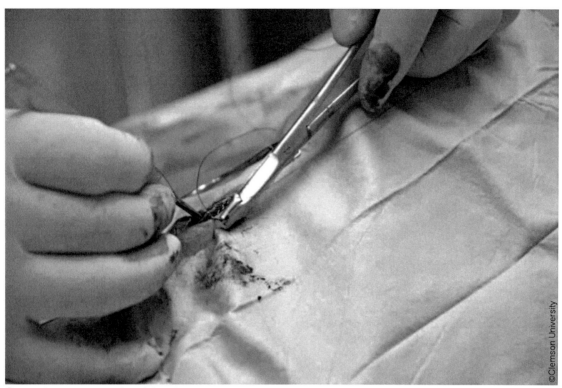

Suturing a surgical incision

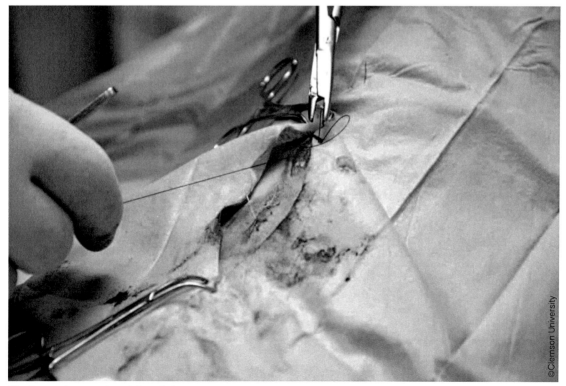

Sutured surgical incision

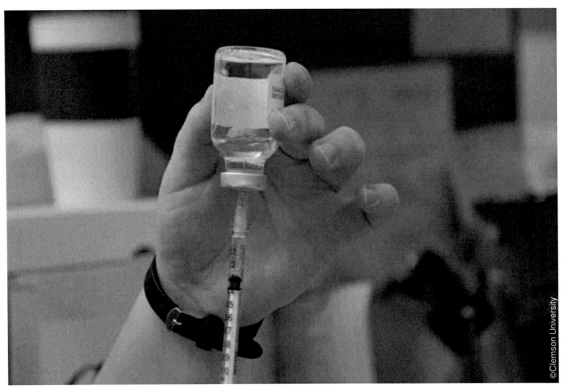

Drawing up medication

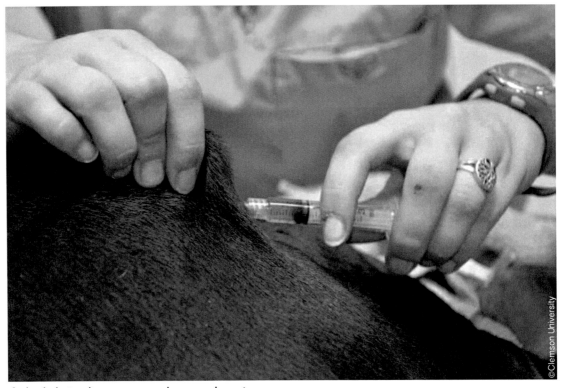

Administering a vaccine, subcutaneous

Glossary

A

A-band - the area between two I bands of a sarcomere, marked by partial overlapping of actin and myosin filaments; does not contract during muscle contraction.

accessory lobe - one of the lobes of the right lung through which the caudal vena cava passes

acetabulum - socket of the hip joint that receives the head of the femur

acromion - the flattened triangular projection of the scapular spine that serves as the attachment for the deltoid and trapezius muscles

adipocyte - fat cells; the main component of adipose tissue

adluminal compartment - the area of a seminiferous tubule defined by the tight junctions of Sertoli cells at the lower boundary and by the lumen of the seminiferous tubule at the upper boundary

adrenal gland - endocrine gland located above the kidney, chiefly responsible for releasing hormones in response to stress through the synthesis of corticosteroids such as cortisol and catecholamines

alveoli - *Respiratory*: The terminal end of a bronchioles and the principal site of gas exchange between the lungs and the blood. *Reproduction*: the milk secreting unit of the mammary gland

anorectal junction - the transition from the rectum to the anal canal

aorta - the largest artery in the body that leaves the left ventricle of the heart and through which all systemic, oxygenated blood passes

apex - the ventrocaudal region of the heart

aponeurosis – a flat, broad tendon that binds muscles together or connects muscle to bone

arrector pili muscle - small muscle that attaches to the base of a hair follicle at one end and to dermal tissue on the other end

artery - *Systemic*: carries oxygenated blood from the left ventricle to the body. *Pulmonary*: carries oxygen-depleted blood from the right ventricle to the lungs

atrioventricular valve - small valves that prevent backflow from the ventricles into the atrium

atrium - an entry chamber of the heart that leads into the ventricles

auditory ossicles - three bones (malleus, incus, and stapes) in the middle ear that are involved in the transmission of sound waves across the tympanic membrane

auricle - the ear-shaped muscular appendage projecting from each atrium of the heart

autoclave - device used for sterilizing instruments and equipment using high pressure and temperature

azygos - vein that runs along the side of the vertebral column and functions in draining the thoracic and abdominal walls

B

biceps brachii - skeletal muscle involved in the movement of the elbow and the shoulder; commonly referred to as the "biceps" muscle

bicornuate uterus - a uterus consisting of distinct uterine horns (cornua)

bifurcation - the division of a main body into two parts

bladder - the organ that collects urine excreted by the kidneys before disposal by urination; urine enters the bladder via the ureters and exits via the urethra

Bowman's capsule - a cup-like sac at the beginning of the tubular component of a nephron that performs the first step in the filtration of blood to form urine

brachial - pertaining to the forelimb

brachycephalic - in skull anatomy, having a short facial area and a wide, globular cranium; includes the English bulldog and the Pekingese

bronchus - a passage airway in the respiratory tract that conducts air into the lungs

bulbourethral gland - a small exocrine gland present in the male reproductive system that produced a viscous secretion prior to ejaculation; also known as the cowper's gland

bulla - a fluid-filled blister

C

canine - dog

caput epididymis - head of epididymis

cardiac muscle - involuntary striated muscle of the heart

carotid artery - artery that supplies the neck and the head with oxygenated blood

cartilage - firm, flexible connective tissue that cushions bones at joints; also located in the ears, nose, and trachea

cauda epididymis - tail of epididymis

caudal - rear or tail region; pertaining to the hind parts

caudal vertebrae - tail bones

cecum - a pouch, usually peritoneal, that is considered to be the beginning of the large intestine. It receives fecal material from the ileum and connects to the ascending colon of the large intestine

cephalic - *General Anatomy:* relating to the head. *Circulatory:* the cephalic vein is located on the cranial aspect of the forelimb and is one of the most common peripheral veins used for blood collection

cerebellum - region of the brain that plays an important role in movement, balance, and equilibrium

cerebrum - the largest region of the brain containing lobes with specific function including movement, recognition, and visual processing

cervical lymph node - lymph node lying near the surface of the neck

cervical vertebrae - vertebrae located directly below the skull

cervix - the narrow, necklike passage forming the lower portion of the uterus where it joins with the anterior wall of the vagina

chondrocytes - cartilage cells

chordae tendonae - cord-like muscles that connect the atrioventricular valve to the heart wall

circulatory system - organ system transporting blood throughout the body

clavotrapezius muscle - (feline only) a muscle in the trapezius group that arises from the back of the skull and acts to extend the humerus in a forward direction

cochlea - spiral-shaped auditory portion of the inner ear that converts sound impulses into nerve transmissions

collecting duct - part of a series of tubules and ducts that connect the nephrons to the ureter and participates in electrolyte and fluid balance through reabsorption and excretion

colloid - a thick fluid produced and located in the follicles of the thyroid that stores hormones

colon - the part of the large intestine extending from the cecum to the rectum that forms fecal matter

convoluted tubules - tubules in the kidney that carry fluid, absorb nutrients, ions, proteins, vitamins, and water

cornea - the outer, transparent layer of the eye covering the iris and the pupil

coronoid process (mandible) - the anterior projection of the mandibular ramus that serves as the insertion point for the temporalis muscle

corpus albicans - a white scar-like fibrous ovarian structure resulting from the regressed form of the corpus luteum

corpus cavernosum - the cavernous erectile tissue in the central portion of the penis that allows for influx of blood during erection of the penis

corpus epididymis - body of epididymis

corpus luteum - the endocrine structure on the ovary that is formed after ovulation and produces progesterone

corpus spongiosum - the portion of erectile tissue in the penis that surrounds the penile urethra

cortex - the outermost layer of an organ

costal cartilage - segments of hyaline cartilage that articulate with the sternum and rib cage, and contribute to the elasticity of the thoracic walls during respiration

croup - the hindquarters or rump of a quadruped

cuterebra ("wolf") worm - parasitic larva of the cuterebra fly that enter a host through the mouth, nose, or an open wound; the larva forms a hole in the host's skin through which they breathe

cranial - pertaining to the skull or the cranium

cremaster muscle - a striated muscle continuous with the internal oblique muscle that partially surrounds the spermatic cord and attaches to the parietal vaginal tunic

D

deciduous teeth - the first set of teeth; also called "milk teeth" or "baby teeth"

deltoid muscle - covers the shoulder and is used to extend the forelimb

dermis - thick layer of tissue located between the epidermis and subcutaneous tissue composed of connective tissue, blood and lymph vessels, sweat glands, nerve endings, and hair follicles

dewclaw - a functionless vestigial claw on the paw that does not reach the ground when walking

diaphragm - dome shaped skeletal muscle associated with breathing that forms a partition between the thoracic and abdominopelvic cavities

diaphysis - the midsection or shaft of a long bone

digastric muscle - muscle with an insertion point on the ventromedial border of the mandible; acts to open the jaws

digestive system - a group of organs working together to convert food into energy and basic nutrients for the entire body

digital extensor - tendon that runs down the front of the leg and provides stability to the fetlock, pastern, and digit joints

digital flexor - tendons of the lower leg

distal - directional term describing the point furthest from the body

dolichocephalic - in skull anatomy, having a relatively long face length compared to the width of the head; examples include the Greyhound and the Collie

dorsal - directional term referring to the back

duodenum - the first section of the small intestine

dystocia - difficult birth

E

endocrine system - the system of glands, each of which secretes different hormones that regulate metabolism, growth and development, tissue function, sexual function, reproduction, sleep and mood

endometrium - the mucosal lining of the uterus

epididymis - a single, narrow, tightly-coiled tube connecting the efferent ducts from the rear of each testicle to the ductus deferens. It serves as a transport, storage, and maturation site for spermatozoa and includes caput, corpus, and caudal regions

epiglottis - a flap of elastic cartilage situated at the base of the tongue that covers the opening to the air passages when swallowing

epiphysis - the rounded end of a long bone

epithelial cell - cells that line cavities and cover flat surfaces; these may be single or multiple layers

esophagus - muscular tube of the gastrointestinal tract through which food passes from the pharynx to the stomach

estrus - the period of sexual receptivity in the female; also referred to as "heat"

eustachian tube - connects the tympanic cavity with the nasopharnx and allow the equalization of the pressures on the two sides of the eardrums; also called "auditory tube"

external abdominal obliques ("external oblique") - abdominal muscle that functions in pulling the chest downwards to compress the abdominal cavity

F

fascicle - a bundle of muscle fibers

feline - cat

femoral - related to the leg or femur

fetus - the unborn, developing offspring of an animal; the term applies from the embryo stage until parturition

filament - microscopic thread-like structures that make up striated muscle fibers, also called myofilaments

filiform papillae - the most numerous papillae that are arranged in regular rows running parallel to the median line of the tongue; also known as "conical papillae"

flea - hematophagic, wingless external parasite noted for its ability to leap

frontalis muscle - cutaneous muscle stretching over the forehead into the upper eyelid

fungiform papillae - epithelial taste organs on the surface of the tongue embedded with taste buds on the superior surface

G

gallbladder - a pear-shaped organ located below the liver that stores bile produced by the liver

ganglion - a nerve cell whose body is located outside of the central nervous system

gastrocnemius - the largest, most superficial muscle of the lower leg, extending from above the hock to the hoof

germinal center - the area in the center of a lymph nodule containing aggregations of actively proliferating lymphocytes

germinal epithelium - the epithelium of the seminiferous tubule that produces spermatozoa

gingiva - the mucous membrane and supporting fibrous tissue that immediately surrounds the teeth; also known as "gums"

gingivitis - inflammation of the gingiva, or gum tissue

glabrous - smooth, hairless skin, as in the paw pads and nose

glomerulus - a network of capillaries that performs the first step of filtering blood at the beginning of the nephron in the kidney

gluteal lymph node - lymph nodes surronding the gluteal muscles

goblet cells - specialized epithelial cells in the digestive system and respiratory system that secrete mucus

gray matter - neural tissue containing nerve cell bodies and dendrites

gracilis - the most superficial medial femoral muscle that assists in drawing the hindlimb inward and bending the stifle

H

H-zone - area within the A-band that is composed of only thick filaments

Hassall's corpuscle - small spherical or ovoid bodies in the medulla of the thymus composed of concentric arrays of epithelial cells

Haversian canals - channels within bone through which blood vessels, nerve fibers, and lymphatics pass

heartworm - parasitic roundworm transmitted by mosquitos that lives in the heart and pulmonary arteries of an animal

hock - joint between the tarsal bones and the tibia

humerus - long bone of the forelimb that articulates with the ulna and radius

hyperplasia - increase in cell numbers

hypothalamus - the region of the brain involved in coordinating the physiologic responses of the organs that maintain homeostasis

I

ileum - the final section of the small intestine

iliac - pertaining to the ilium or pelvic bones

iliac lymph node - lymph node located below the ilium

ilium - the dorsal, upper, and largest of the three bones of either lateral half of the pelvis that joins with the ischium and pubis to form part of the acetabulum

incisors - the front teeth used for shearing vegetation

inferior - beneath or bottom

inguinal lymph node - lymph nodes located near the vastus lateralis muscle of the hindlimb

integumentary system - the organ system that provides covering of the body and protects it from water loss and abrasion, comprising the skin and its appendages

intubation - the insertion of a flexible plastic tube into the trachea to maintain an open airway during surgical procedures

iris - the colored part of the eye surrounding the pupil that regulates the amount of light entering the eye

ischium - bone that makes up the lower and back portion of the hip bone

Islets of Langerhans - isolated groups of pancreatic cells that produce insulin and glucagon

isthmus - *General Anatomy:* a narrow passage connecting two separate structures. *Female Reproduction:* connecting the ampulla of the oviduct to the uterus

J

jejunum - the middle section of the small intestine between the duodenum and the ileum

jugular vein - vein that brings deoxygenated blood from the head back to the heart

K

kidney - an organ that filters blood, regulates electrolytes, maintains acid-base balance, and regulates blood pressure. It also controls the body's fluid balance and produces urine

L

lacunae - small cavities within bone that contain osteocytes

lamina propria - highly vascular connective tissue and a constituent of the mucous membranes lining the gastrointestinal tract

large intestine - the final region of the digestive system that functions to absorb water from indigestible food matter and to pass waste material from the body

laryngoscope - a tubular endoscope inserted through the mouth; it is used to examine the interior of the larynx and assist in intubation

lateral - directional term used to describe the side

lateral digital extensor - tendon that runs down the front of the leg and extends the carpal, pastern, and coffin joints

latissimus dorsi - large, flat muscle responsible for the extension, rotation, and flexion of the shoulder joint

levator labii - a facial muscle arising from the lower margin of the orbit and inserting into the muscular substance of the upper lip, which it elevates

levator nasolabialis - a muscle originating on the surface of the skull in front of the eye that lifts the upper lip and dilates the nostril or wrinkles the snout

Leydig cells - cells found in the interstitial compartment of the testis that produce testosterone

ligament - band of connective tissue that connects two bones (or cartilage) at a joint

limbus - the distinctive marginal region of the cornea of the eye that is continuous with the sclera

linea alba - a median tendinous line on the abdomen formed of fibers from the aponeuroses of abdominal muscles

liver - glandular vascular organ that secretes bile, stores and filters blood, synthesizes proteins, and is involved in metabolic functions

Loop of Henle - the portion of a nephron that leads from the proximal convoluted tubule to the distal convoluted tubule. Its main function is to create a concentration gradient in the medulla of the kidney

lumbar - the third major region of the spine below the cervical spine and thoracic spine

lumen - the inside space or cavity of a tubular structure, such as an artery or intestine

lung - the essential respiration organ that functions to transport oxygen and carbon dioxide

lymph node - an oval body in the lymphatic system that produces and houses lymphocytes and filters microorganisms from lymph

lymphatic system - network of vessels that transport fluid, fats, proteins, and lymphocytes to the bloodstream as lymph, removing microorganisms from tissues

M

mandible - lower jaw bone

mammary parenchyma - fully developed epithelial or glandular tissue of the mammary gland

masseter - a jaw muscle that runs between the cheekbone and the lower jawbone and is used for chewing (mastication)

mast cell - tissue-bound cell that mediates inflammatory responses such as allergic reactions; releases granules of inflammatory biochemicals, such as histamine, against invading microorganisms

mature follicle - follicle that is ready to ovulate; often called the Graafian follicle

maxilla - the upper jaw bone

M-Band - the narrow band in the center of the H-zone of a sarcomere

medulla - *Lymphatics*: central portion of a lymph node. *Urinary*: the innermost region of the kidney

membrane nictitans - a transparent inner (third) eyelid that protects and moistens the eye

mesaticephalic - in skull anatomy, having relatively harmonious proportions between the length of the skull and its width; includes the Golden Retriever and the Beagle

middle gluteal - the middle of three gluteal muscles situated on the outer surface of the pelvis that functions to move the thigh

molar - posterior teeth of the mouth that grind feed

muzzle - *General Anatomy*: the projecting part of an animal's face, including the jaws, mouth, and nose; snout. *Veterinary Techniques*: a fitted guard around the mouth of an animal to prevent biting

mylohyoid muscle - a neck muscle that controls the tongue by depressing the mandible and raising the floor of the mouth during the first phase of deglutition

myofibril - a slender thread of muscle fiber, made up of several filaments

myometrium - the smooth muscle layer of the uterus consisting of an inner circular layer and an outer longitudinal layer

N

nasal bone - bridge of the nose

nephron - the basic structural and functional unit of the kidney. Its chief function is to regulate the concentration of water and soluble substances by filtering the blood, reabsorbing what is needed and excreting remaining substances as urine

neuron - nerve cell

nucleus - the control center of the cell that contains genetic material

O

occipital - the unpaired bone constituting the back and part of the base of the skull

omentum - an apron-like, fatty layer of peritoneum surrounding abdominal organs; also referred to as "omental fat"

omohyoid muscle - compresses the hyoid bone (at the base of the mandible) and the larynx

omotransversarius muscle - originates on the wing of the atlas vertebra and terminates on the distal spine of the scapula; pulls the thoracic limb cranially and the head and neck laterally

oocyte - immature ovum located in the ovarian follicles

orbicularis oculi - facial muscle used to close the eyelids

orbicularis oris - grouping of skeletal muscle around the lips that functions in compressing the lips together

osteocyte - star-shaped bone cell located in the lacuna that communicates with other osteocytes and blood vessels in the matrix

osteon - the functional unit of compact bone tissue containing blood vessels, nerves and connective tissue that provides the bone's blood and nerve supply

osteophyte - abnormal bony outgrowth, usually found along a joint; also referred to as a "bone spur"

ovary - either of the two female gonads that produce oocytes and the sex hormones estrogen and progesterone

oviduct - the small, convoluted ducts that transport ova and sperm; the oviduct consists of the ampullary and isthmic junctions

P

patella - a triangular bone situated at the front of the knee that serves to protect the joint

palatine - either of a pair of bones that are situated behind and between the maxilla; forms the inner portion of the hard palate

pampiniform plexus - a specialized group of veins originating in the spermatic cord and terminating on the dorsal pole of the testis that provides countercurrent heat exchange for the testes; consists of the testicular vein and testicular artery

pancreas - a glandular organ in the endocrine and digestive systems that produces hormones in the blood and digestive enzymes to assist in the absorption of nutrients in the small intestine

parathyroid gland - small endocrine glands located in the neck that produce parathyroid hormone (PTH) and regulate the body's blood calcium levels

parenchyma - essential and distinctive tissue of an organ

parotid gland - the major salivary gland located below the ear that excretes saliva to lubricate and soften feed prior to digestion

parturition - to give birth

pastern - a part of the limb (forelimb or hindlimb) extending from the carpal joint to the top of the paw

pelvis - the bone that connects the base of the spine to the upper end of the rear legs

phagocyte - a cell, such as a white blood cell, capable of ingesting microorganisms, cells, debris, and other substances

pinna - visible part of the ear outside the head

pituitary gland - an endocrine gland located in the brain that secretes hormones and regulates homeostatic processes; partitioned structurally and functionally into an anterior and posterior lobe

plaque - colorless film of microbes, organic and inorganic material that forms on the surface of teeth

podocytes - epithelial cells in the glomerular capsule of the kidney that make up the barrier surrounding the capillaries through which blood is filtered

popliteal lymph node - a lymph node found behind the stifle joint between the semimembranosus and semitendinosus muscles

popliteal vein - vein located behind the stifle that carries blood from the stifle joint and muscles in the hindlimb back to the heart

premolar - transitional teeth located between the canine and the molars

primary follicle - consists of a primary oocyte with a single layer of cuboidal/columnar follicular cells

prostate - male accessory sex gland that secretes alkaline fluid

proximal - term used to describe the point closest to the attachment of the body

proximal convoluted tubule - the portion of the duct system of the nephron of the kidney which leads from Bowman's capsule to the Loop of Henle

pupil - the opening of the iris

Purkinje fibers - specialized cardiac muscle fibers that facilitate coordinated contraction and rapid impulse transmission from the atrioventricular node to the ventricles

Q

quick - the thickened epidermis beneath the free distal end of the claw of a digit; formally referred to as the "hyponychium"

R

radius - the shorter bone of the forelimb lying parallel to the ulna that forms the elbow joint on the proximal end and the carpal joint on the distal end

rectum - the part of the descending colon located within the pelvis that serves to store feces prior to expulsion

renal - pertaining to the kidneys

renal cortex - the outer portion of the kidney between the renal capsule and the renal medulla

renal medulla - the innermost region of the kidney

reproductive system - a group of organized structures responsible for development of new life and the continuation of a species

respiratory system - a group of specific organs and structures used for the process of respiration

retractor penis - a pair of smooth muscles originating on the ventral surface of the first few caudal vertebrae and attaching on the lateral and urethral surfaces of the penis; relaxation of this muscle is required for full penile protrusion and erection

retropharyngeal lymph node - one of three groups of lymph nodes located between pharynx and prevertebral layer of cervical fascia; they receive lymph from the nasopharynx, the auditory tube, and the atlantooccipital and atlantoaxial joints

S

sacral lymph node - lymph nodes located in the caudal region of the spine and situated in the sacrum

sacrum - the large bone at the base of the spine made of fused sacral vertebrae and located between the lumbar vertebrae and the coccyx; triangular in shape and makes up the back wall of the pelvis

salivary gland - exocrine organs that secrete saliva into the oral cavity

sarcomere - the functional unit of skeletal muscle, the segment between two Z-lines

sartorius muscle - superficial muscle of the thigh that inserts at the cranial border of the tibia; acts to flex the hip and stifle while the limb is protracting and to contribute stifle extension during standing

scapha - the large flat concave internal side of the pinna (ear)

scapula - shoulder blade; flat, asymmetrical bone that joins the forelimb to the trunk

sciatic - nerve originating in the caudal region of the spine that innervates the quadriceps muscle of the thigh and the deep gluteal muscle

sclera - the dense outer coating of the eyeball that forms the white of the eye

sebaceous gland - gland that expels an oily secretion into the hair follicle

secondary follicle - an ovarian follicle characterized by having two or more cell layers that begin to secrete follicular fluid

semicircular canal - one of three fluid filled ducts, designated anterior, posterior, and lateral, in the inner ear that sense movement and changes in equilibrium

seminiferous tubules - the tightly wound tubules within the testes that produce spermatozoa

semitendinosus muscle - superficial caudal musculature of the pelvic limb

Sertoli cells - somatic cells in the seminiferous epithelium that initiate spermatogenesis

skeletal muscle - striated, voluntary tissue stimulated by somatic nervous system

skeletal system - all of the bones and cartilage of the body that provide supporting framework, produce blood cells, and store minerals

small intestine - the major site of nutrient absorption in the digestive tract comprised of three regions: the duodenum, jejunum, and ileum

spermatic cord - a cord-like cllection of tissues containing the testicular artery and vein, lymphatics, pampiniform plexus, nerves, the cremaster muscle, and the ductus (vas) deferens

spinal cord - a long, thin bundle of nervous tissue that extends from the brain; responsible for neural transmissions between the central nervous system and the peripheral nervous system

spleen - a large lymphoid organ that acts as a reservoir for blood cells, removes and forms blood cells as a blood filter, and produces lymphocytes to aid in fighting infection

spleen pulp - splenic tissue

sternomastoid muscle - muscle that flexes the vertebral column and rotates the head

sternum - an elongated bone in the center of the chest that is the site of attachment for the first seven pairs of ribs

stifle - the joint between the femur and the tibia corresponding to the human knee

stop - an indentation above the eyes and between the muzzle and braincase (forehead)

striations - the lines in skeletal muscle due to alternating actin and myosin bands in the sarcomere

subcutaneous - beneath the skin

subcutaneous fat - most widely distributed layer of subcutaneous tissue made up of adipocytes

subgingival curette - dental tool used to clean plaque and calculus from the tooth root surface

submandibular lymph node - one of a cluster of lymph nodes grouped around the facial vein at the angle of the jaw

superficial abdominal lymph node - lymph nodes in the abdomen lying near the surface

superior - above or top

tartar - a hard deposit of mostly organic material that forms on teeth at the gum line and contributes to dental decay

teat - projections from the mammary gland from where milk is discharged

temporal - bone that supports the temple on the side of the skull

tendon - an inelastic band of tough fibrous connective tissue that connects muscle and bone

testis - one of the two male gonads

thalamus - gray matter lying beneath the cerebral hemispheres in the brain that relay sensory information from the brain stem to the cerebral cortex

thoracic - referring to the spine, the second segment of the vertebral column after the cervical region

thoracic cavity - a chamber of the body protected by the thoracic wall that includes the heart, lungs, and ribs

thoracic vertebrae - vertebrae between the cervical and lumbar vertebrae where the ribs attach, forming the posterior wall of the thorax

thymus - an organ, located at the base of the neck, that is involved in development of cells of the immune system, particularly T cells

thyroid - a highly vascular, ductless gland located in the anterior neck in front of the trachea. It consists of a left and right lobe connected by an isthmus and functions in the synthesis and storage of thyroid hormones and the regulation of metabolism and calcium levels

tonsil (palatine) - a small mass of lymphoid tissue on either side of the pharynx aiding in immune function

trachea - the cartilaginous tube descending from the pharynx and branching into the left and right bronchi of the lungs, allowing for the passage of air

tragus - the cartilaginous projection anterior to the external opening of the ear

trapezius muscle - muscle of the thoracic limb that fixes the shoulder and lifts, abducts, and draws the limb forward

triceps brachii - large muscle on the back of the forelimb responsible for movement of the elbow joint

tunica albuginea - a dense, white connective tissue covering an organ, such as the testis

tympanic bulla - a hollow, bony chamber that houses the middle ear at the base of the skull

ulna - the longer of the two bones of the forelimb that lies parallel to the radius, and with it forms the elbow joint and the carpal joint

ureter - tubes made of smooth muscle fibers that transport urine from the kidneys to the urinary bladder

urethra - a tube that connects the kidneys to the urinary bladder for the removal of fluids from the body

urinary system - consists of the two kidneys, ureters, the bladder and the urethra; also known as the renal system

uterine body - the part of the uterus above the isthmus, comprising about two thirds of the non-pregnant organ

uterine horn - region of the uterus that connects the oviduct and uterine body; also functions as the site of fetal development

vagina - the female copulatory organ that connects the external genitalia to the cervix; also functions as a birth canal

vas deferens - connects the tail of the epididymis to the ampulla and transports sperm to the pelvic urethra

vein - *Systemic*: carries oxygen-depleted blood to the right atrium of the heart. *Pulmonary*: carries oxygen-ated blood from the lungs to the left atrium of the heart

vena cava - either of the two venae cavae (inferior or superior) that carry deoxygenated blood from the body back to the heart

venipuncture - the puncturing of a vein for medical purposes, e.g. to take blood, to provide fluids intravenously, or to administer a drug

ventral - the directional term referring to the underside of the trunk or belly

ventricle - inferior chambers of the heart that pump blood into pulmonary or systemic circulation

vertebral foramen - a circular opening in the vertebrae through which the spinal cord passes

vulva - the external reproductive genitalia of the female

W

white matter - neural tissue containing nerve fibers with myelin sheath

withers - ridge between the shoulder blades

Z

Z-line - sections muscle fibers into sarcomeres; location where thin myofilaments attach

zonary placenta - a placenta in which chorionic villi attach to the uterus in a well-defined zone or band

zygomatic - referring to the area of the cheek

Please match each of the canine head morphology illustrations to their title.

A. Brachycephalic
B. Dolichocephalic
C. Mesaticephalic

1.

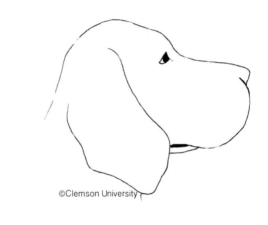
©Clemson University

2.

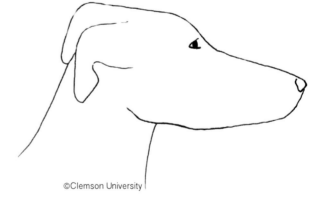
©Clemson University

3.
©Clemson University

Please label feline general anatomy.

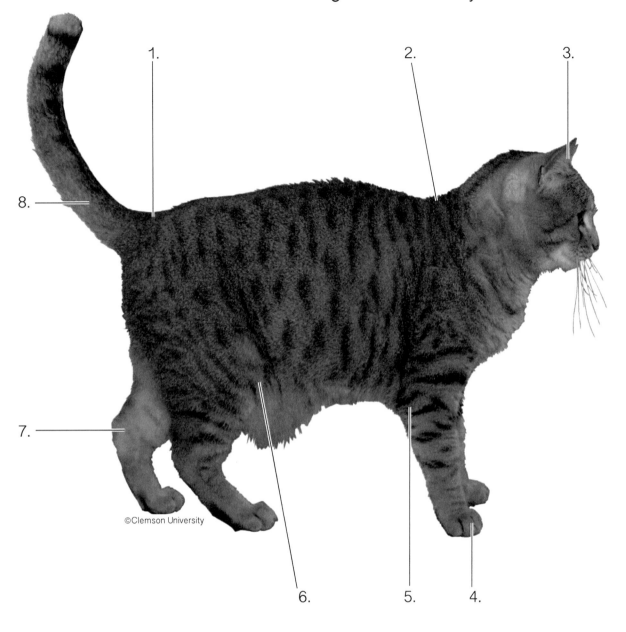

©Clemson University

1. _____

2. _____

3. _____

4. _____

5. _____

6. _____

7. _____

8. _____

Please label the canine directional terms.

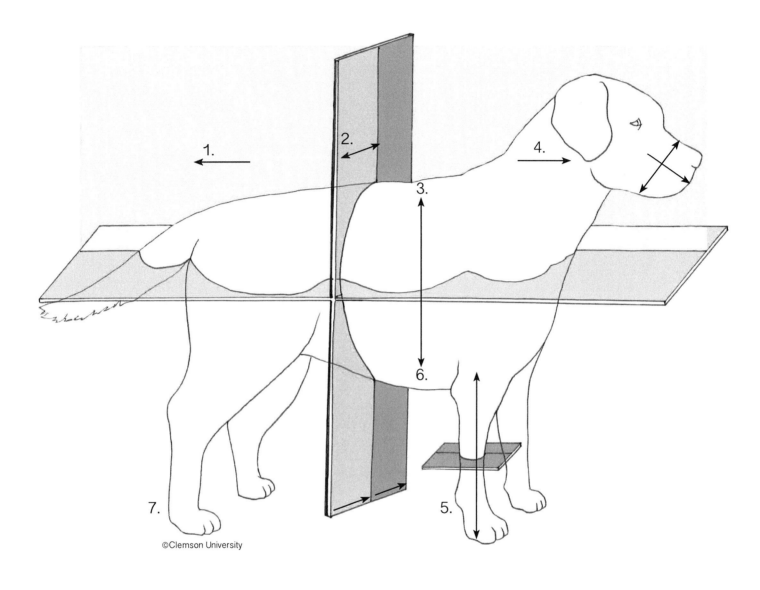

©Clemson University

1. _____

2. _____

3. _____

4. _____

5. _____

6. _____

7. _____

Please label the feline dorsal vertebral regions.

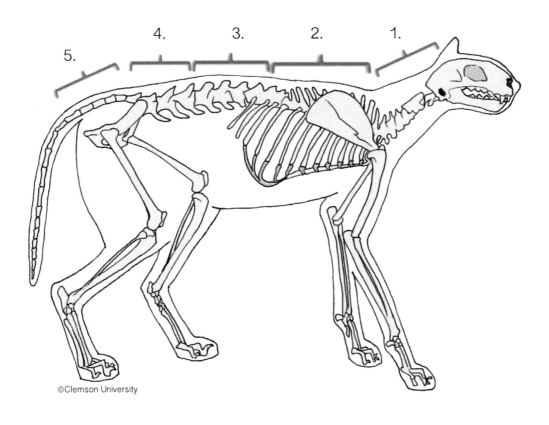

©Clemson University

1. _____

2. _____

3. _____

4. _____

5. _____

Please label the canine skeleton.

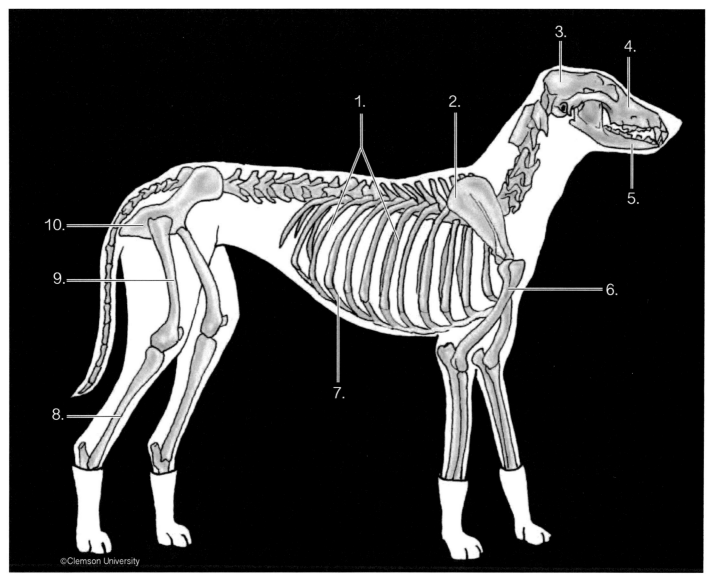

©Clemson University

1. _____

2. _____

3. _____

4. _____

5. _____

6. _____

7. _____

8. _____

9. _____

10. _____

Please label the canine skin.

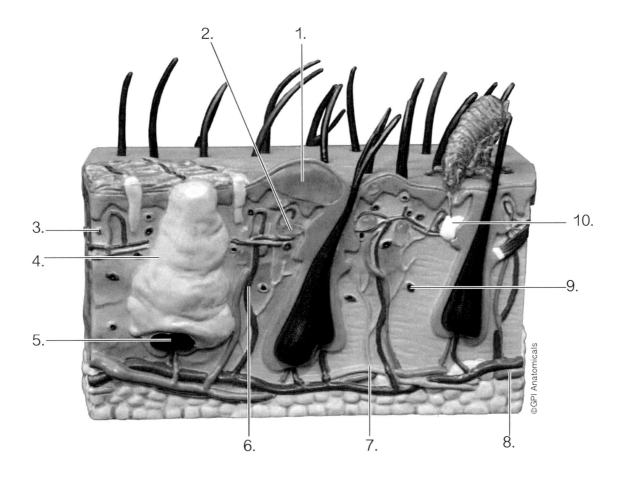

1. _____

2. _____

3. _____

4. _____

5. _____

6. _____

7. _____

8. _____

9. _____

10. _____

Please label the canine paw.

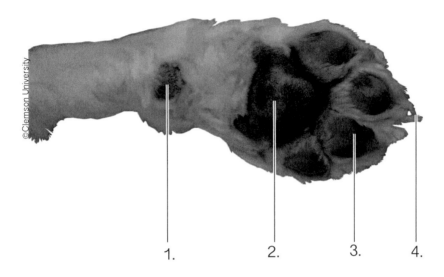

1.   2.   3.   4.

1. _____

2. _____

3. _____

4. _____

Please label the canine foot.

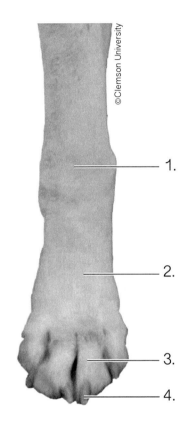

1.

2.

3.

4.

1. _____

2. _____

3. _____

4. _____

Please label the bones of the feline skull.

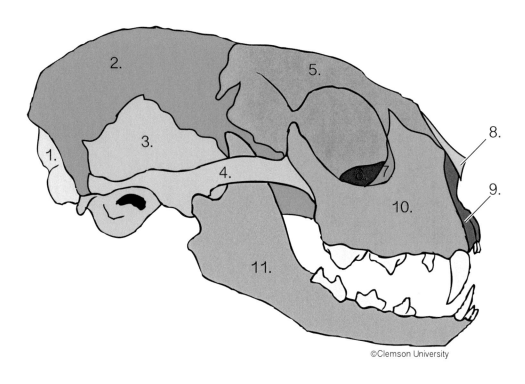

©Clemson University

1. _____

2. _____

3. _____

4. _____

5. _____

6. _____

7. _____

8. _____

9. _____

10. _____

11. _____

Please label the canine skeleton.

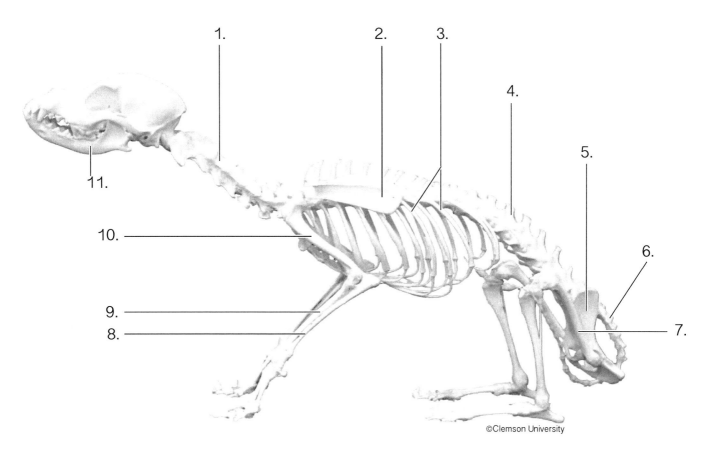

©Clemson University

1. _____

2. _____

3. _____

4. _____

5. _____

6. _____

7. _____

8. _____

9. _____

10. _____

11. _____

Please label the feline skeleton.

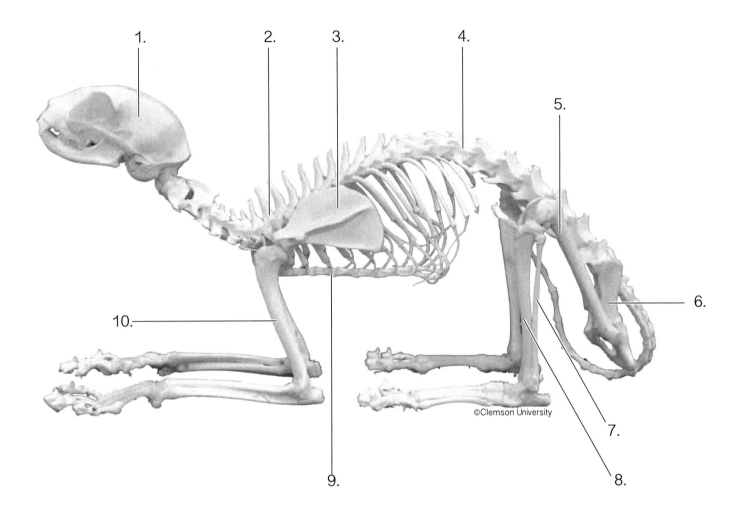

©Clemson University

1. _____

2. _____

3. _____

4. _____

5. _____

6. _____

7. _____

8. _____

9. _____

10. _____

Please label the feline elbow and shoulder.

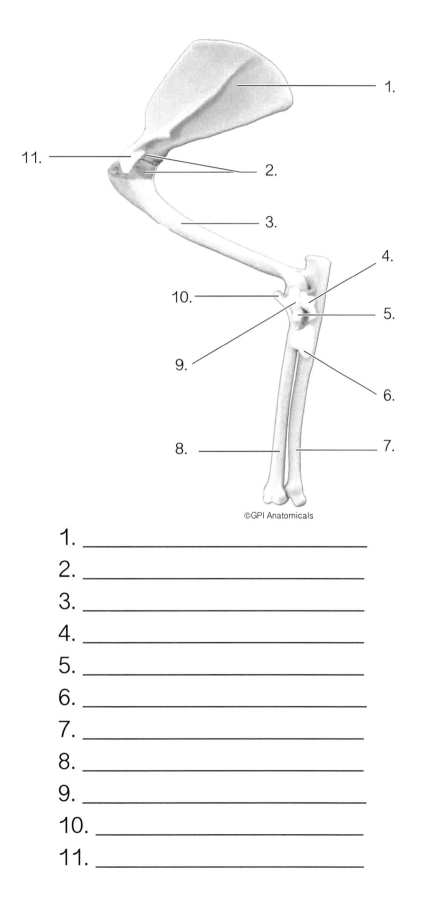

©GPI Anatomicals

1. _____

2. _____

3. _____

4. _____

5. _____

6. _____

7. _____

8. _____

9. _____

10. _____

11. _____

Please label the lumbar vertebrae.

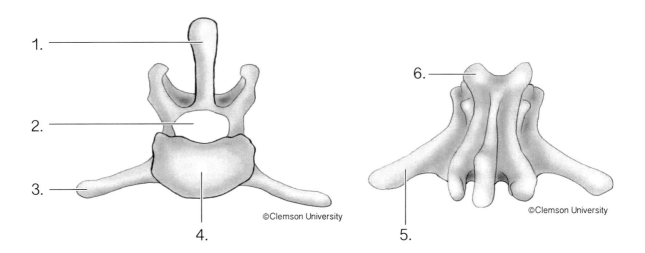

1.

2.

3.

4.

6.

5.

©Clemson University

©Clemson University

1. _____

2. _____

3. _____

4. _____

5. _____

6. _____

Please label the canine vertebral columns.

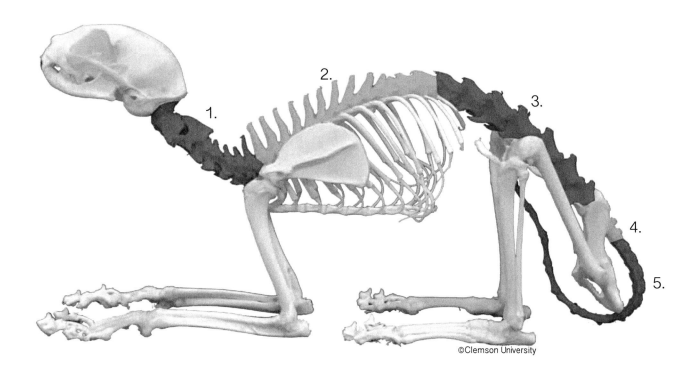

©Clemson University

1. _____

2. _____

3. _____

4. _____

5._____

Please label the feline radiograph.

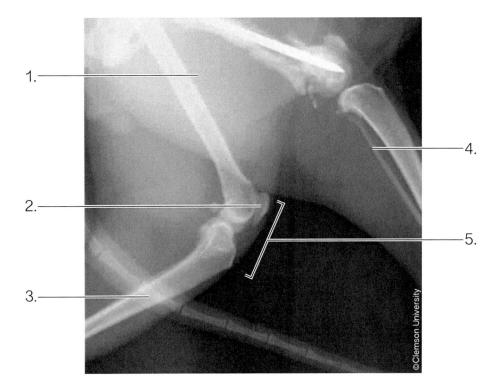

1. _____

2. _____

3. _____

4. _____

5. _____

Please label the skeletal muscle cell.

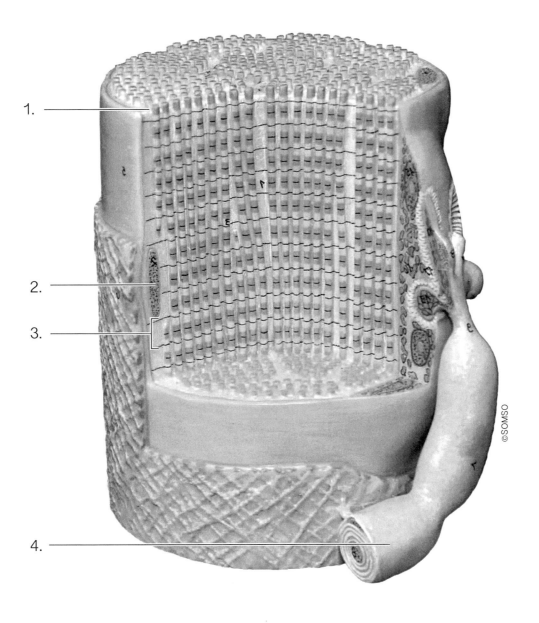

1. _____

2. _____

3. _____

4. _____

Please label the feline cranial muscles.

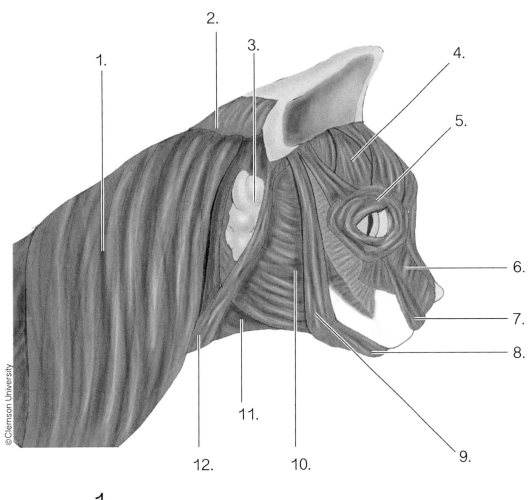

1. _____

2. _____

3. _____

4. _____

5. _____

6. _____

7. _____

8. _____

9. _____

10. _____

11. _____

12. _____

Please label the sarcomere.

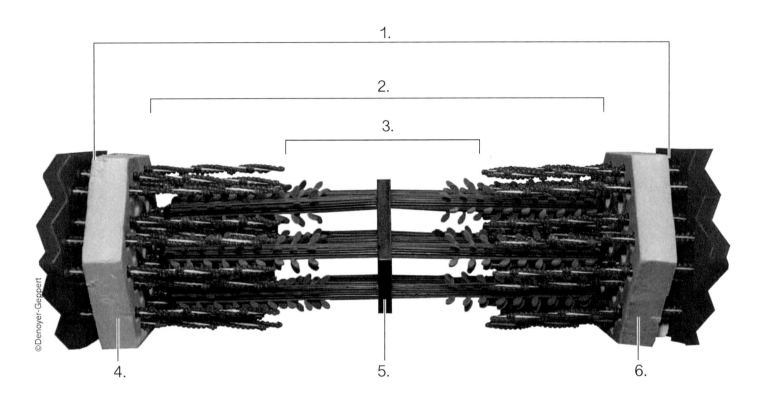

©Denoyer-Geppert

1. _____

2. _____

3. _____

4. _____

5. _____

6. _____

Please label the histological images.

Cervical Spinal Cord

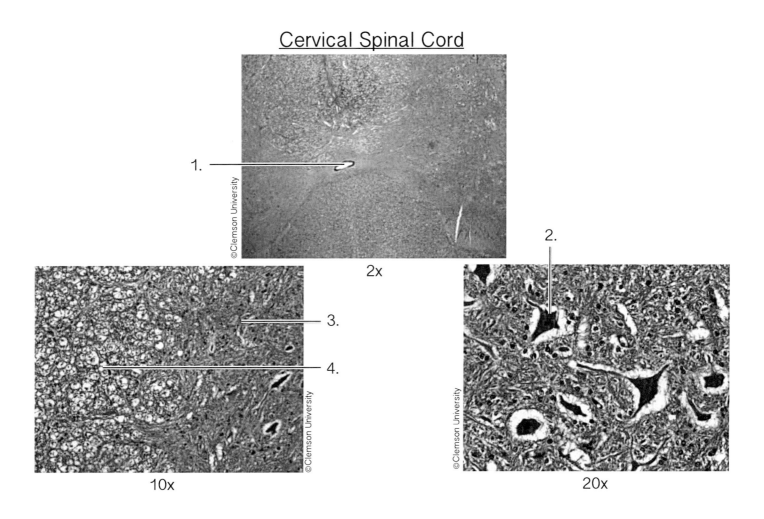

1.

2x

2.

3.

4.

©Clemson University

10x

©Clemson University

20x

1. _____

2. _____

3. _____

4. _____

Please label the canine nerves.

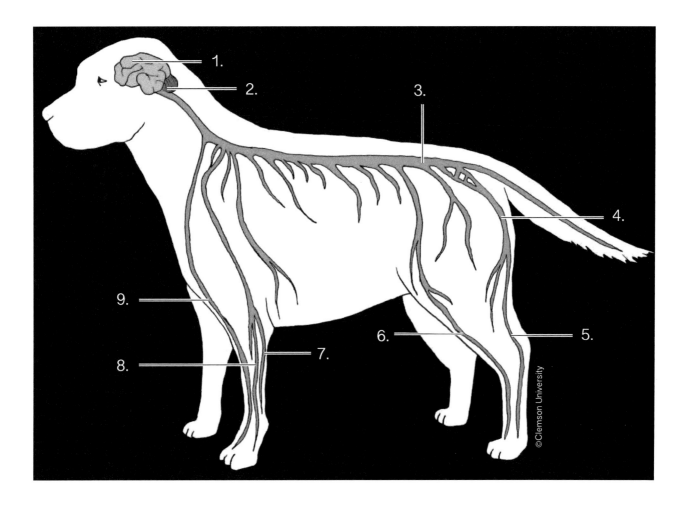

1. _____
2. _____
3. _____
4. _____
5. _____
6. _____
7. _____
8. _____
9. _____

Please label the healthy canine ear.

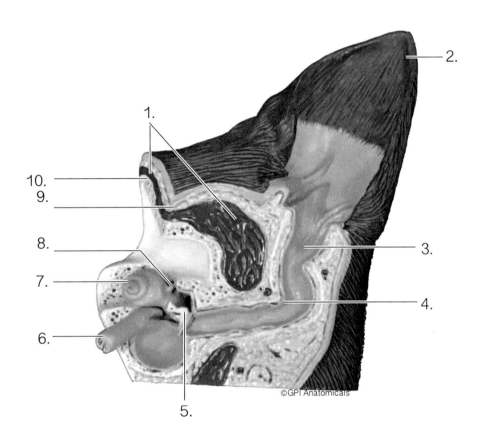

1. _____

2. _____

3. _____

4. _____

5. _____

6. _____

7. _____

8. _____

9. _____

10. _____

Please label the eye diagram.

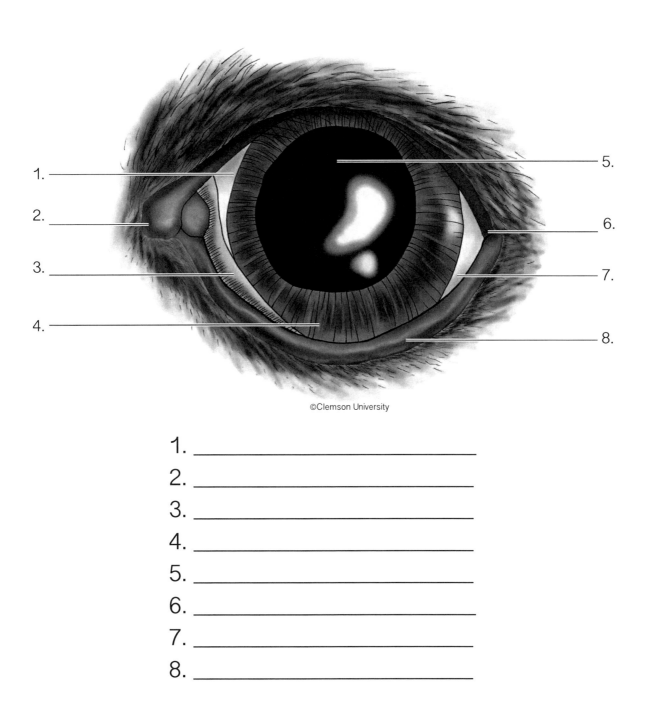

1. _____
2. _____
3. _____
4. _____
5. _____
6. _____
7. _____
8. _____

©Clemson University

1. _____
2. _____
3. _____
4. _____
5. _____
6. _____
7. _____
8. _____

Please label the heart model

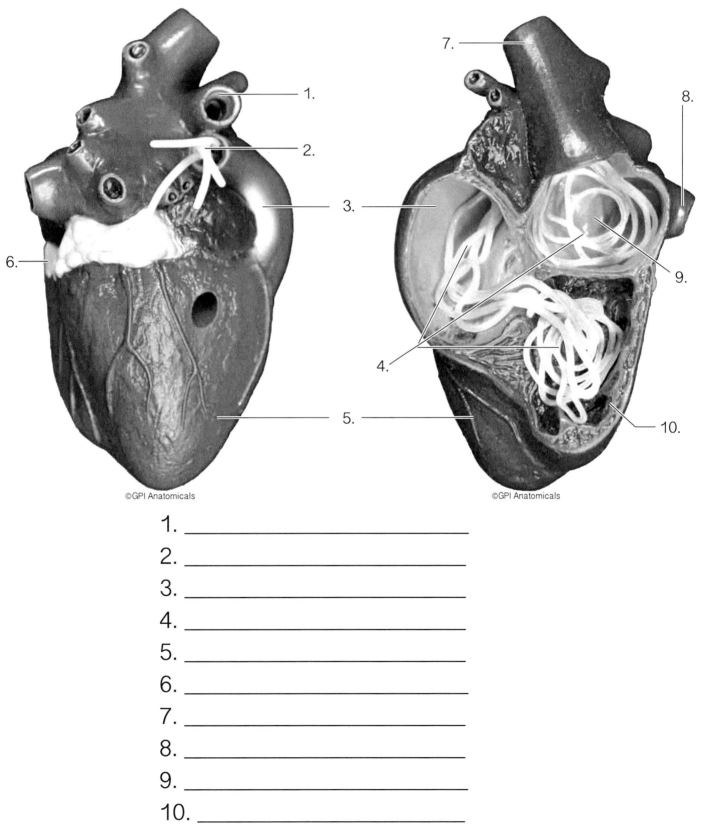

©GPI Anatomicals ©GPI Anatomicals

1. _____

2. _____

3. _____

4. _____

5. _____

6. _____

7. _____

8. _____

9. _____

10. _____

Please label the canine artery diagram.

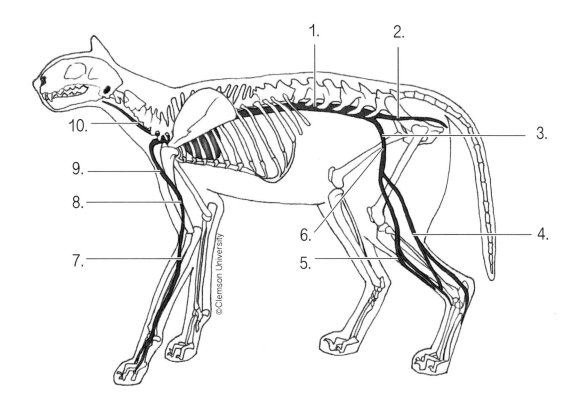

1. _____

2. _____

3. _____

4. _____

5. _____

6. _____

7. _____

8. _____

9. _____

10._____

Please label the feline vein diagram.

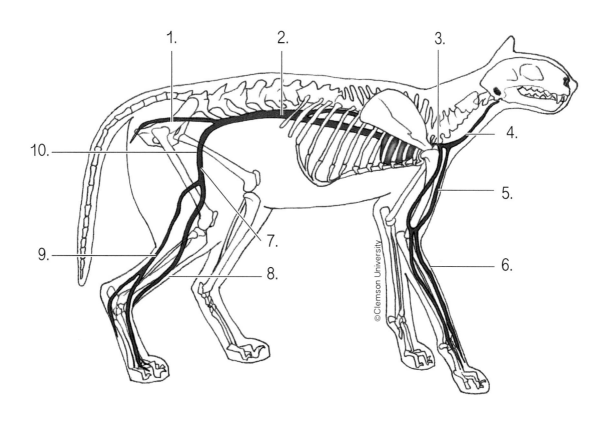

1. _____

2. _____

3. _____

4. _____

5. _____

6. _____

7. _____

8. _____

9. _____

10._____

Please label the canine lymph vessels.

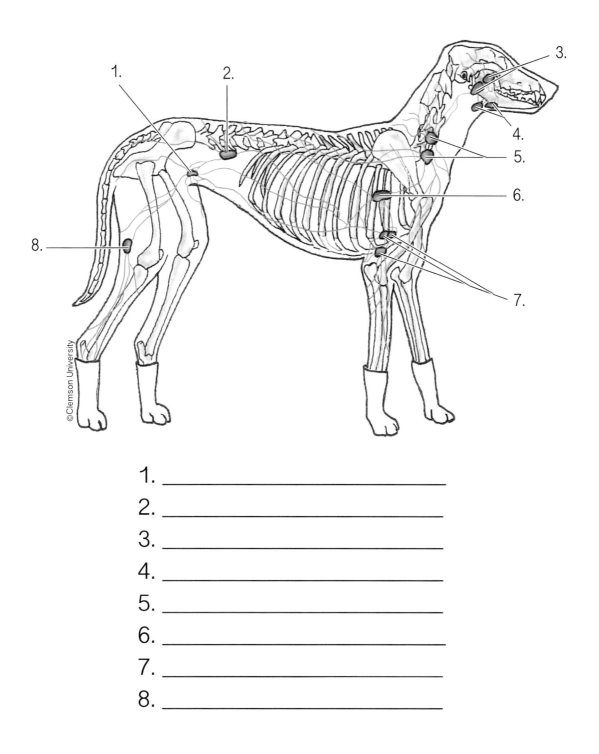

©Clemson University

1. _____

2. _____

3. _____

4. _____

5. _____

6. _____

7. _____

8. _____

Please label the lung diagrams.

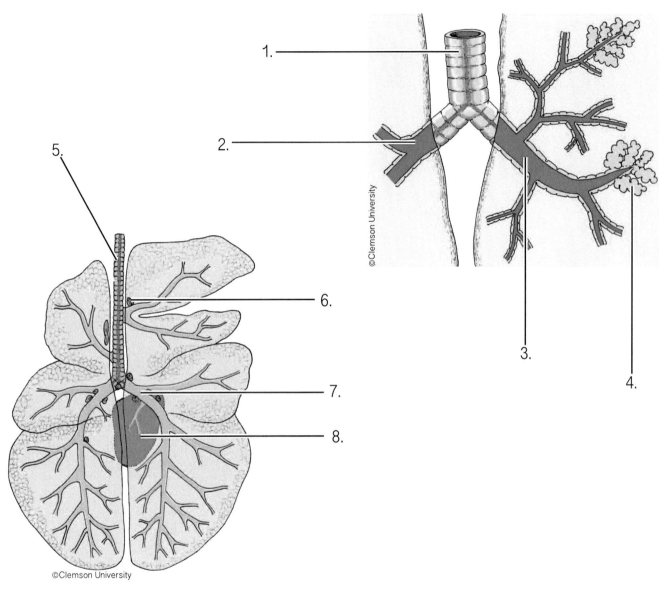

1.

2.

3.

4.

5.

6.

7.

8.

©Clemson University

1. _____

2. _____

3. _____

4. _____

5. _____

6. _____

7. _____

8. _____

Please label the feline lungs.

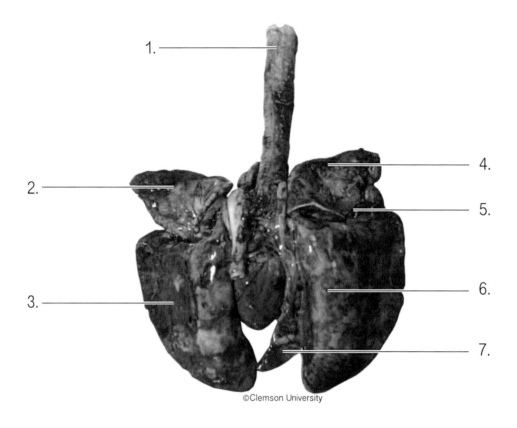

1.

2.

3.

4.

5.

6.

7.

©Clemson University

1. _____

2. _____

3. _____

4. _____

5. _____

6. _____

7. _____

Please label the histological images.

Gut-Associated Lymphoid Tissue

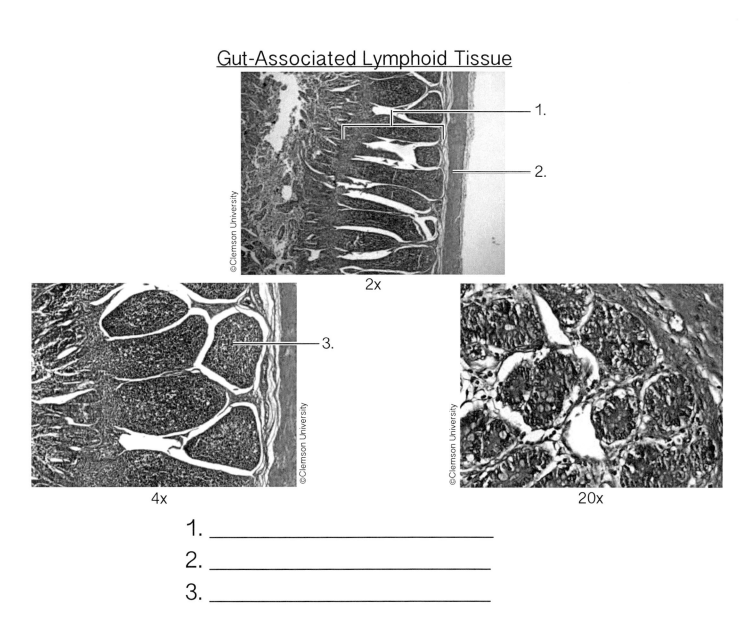

1.

2.

3.

2x

4x

20x

1. _____

2. _____

3. _____

Please label the canine gastrointestinal tract.

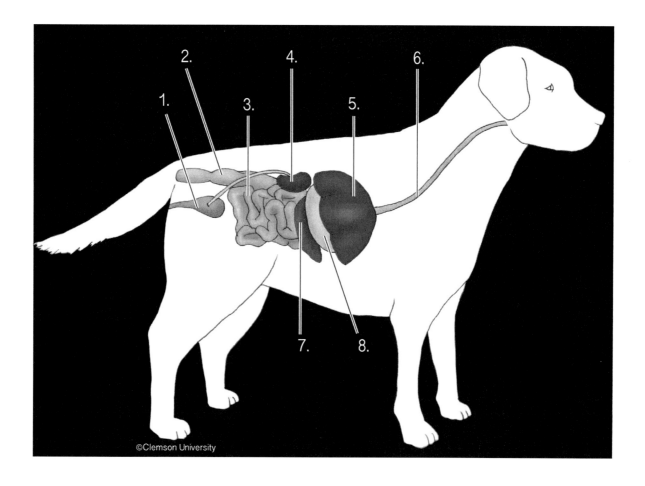

©Clemson University

1. _____

2. _____

3. _____

4. _____

5. _____

6. _____

7. _____

8. _____

Please label the gastrointestinal tract.

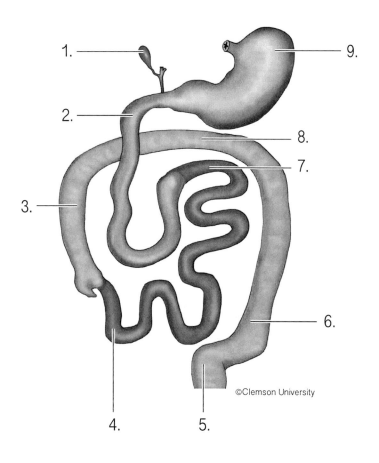

1. _____
2. _____
3. _____
4. _____
5. _____
6. _____
7. _____
8. _____
9. _____

Please label the canine tongue.

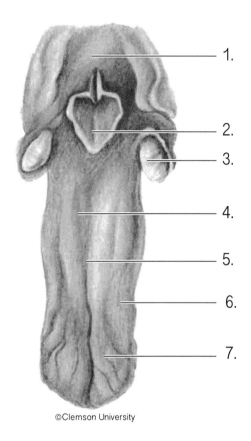

1. _____

2. _____

3. _____

4. _____

5. _____

6. _____

7. _____

©Clemson University

1. _____

2. _____

3. _____

4. _____

5. _____

6. _____

7. _____

Please label the histological images.

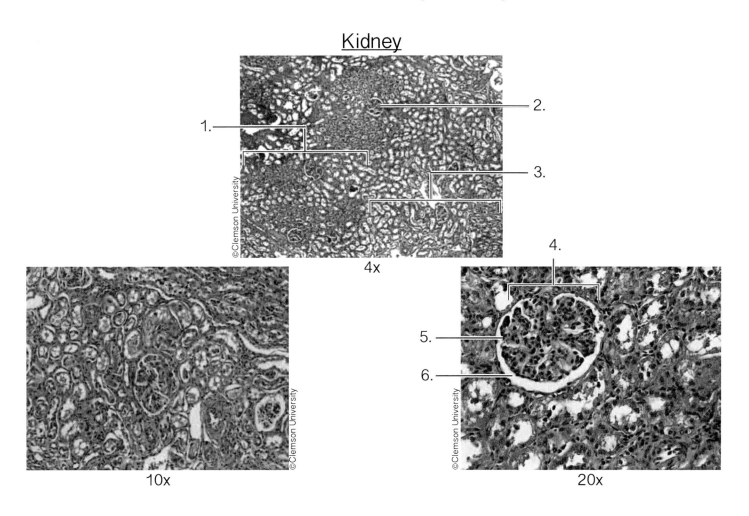

1. _____

2. _____

3. _____

4. _____

5. _____

6. _____

Please label kidney images.

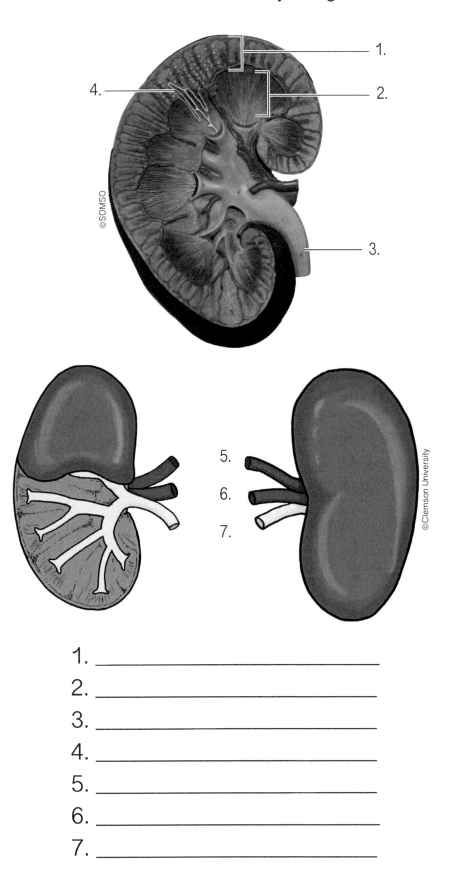

1. _____

2. _____

3. _____

4. _____

5. _____

6. _____

7. _____

Please label the nephron model.

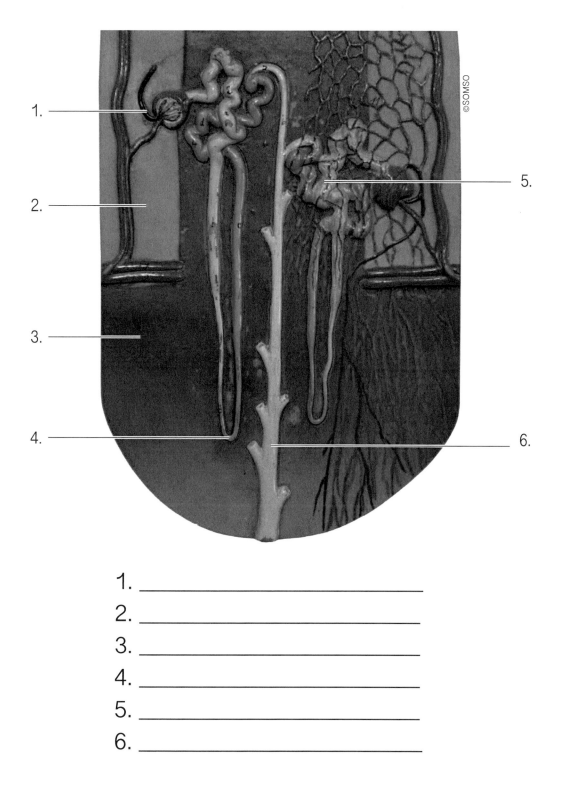

1. _____
2. _____
3. _____
4. _____
5. _____
6. _____

Please label the feline endocrine structures.

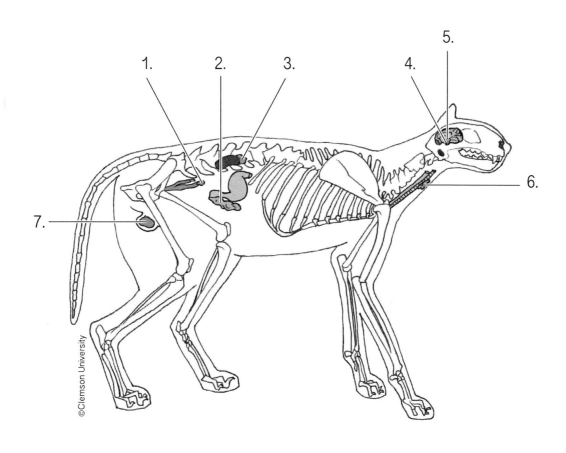

1. _____

2. _____

3. _____

4. _____

5. _____

6. _____

7. _____

Please label the female reproductive organs.

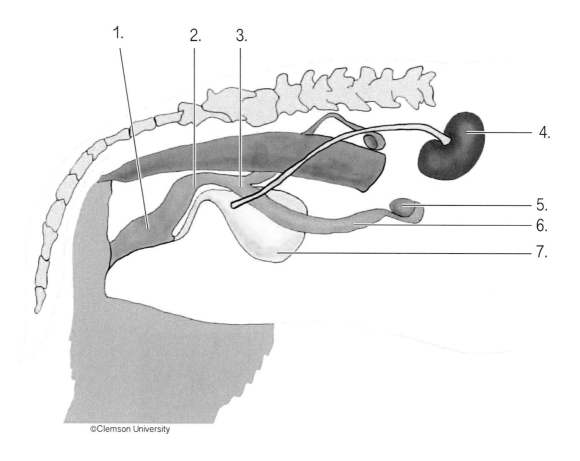

1.
2.
3.
4.
5.
6.
7.

©Clemson University

1. _____

2. _____

3. _____

4. _____

5. _____

6. _____

7. _____

Please label the female reproductive organs.

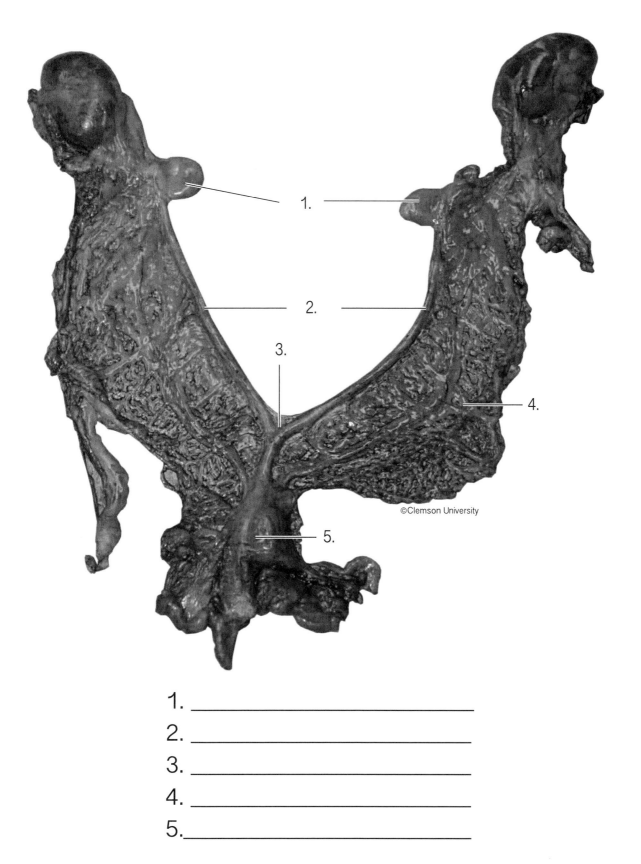

1.

2.

3.

4.

©Clemson University

5.

1. _____

2. _____

3. _____

4. _____

5._____

Please label the ovary.

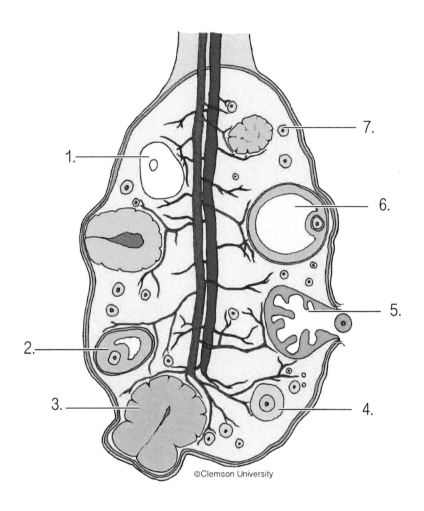

©Clemson University

1. _____

2. _____

3. _____

4. _____

5. _____

6. _____

7. _____

Please label the male canine reproductive organs.

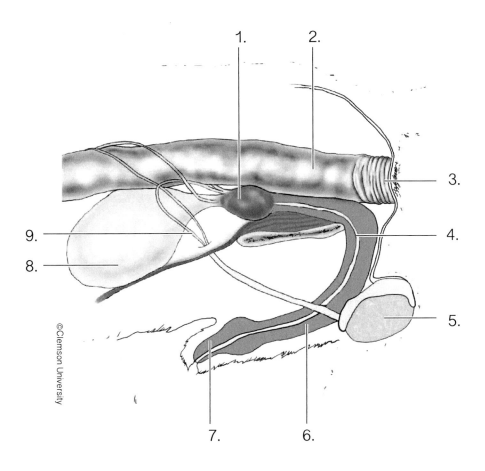

©Clemson University

1. _____

2. _____

3. _____

4. _____

5. _____

6. _____

7. _____

8. _____

9. _____

Made in the USA
Middletown, DE
19 August 2015